T0075220

Good Practices for Disaster Risk Management of Cultural Heritage

This book is a selection of case studies undertaken by cultural heritage and disaster risk management professionals across the world demonstrating good practices for disaster risk management of cultural heritage.

The readers will learn about the practical application of various methodologies, tools, and techniques for disaster risk assessment, mitigation, preparedness, response, and recovery of cultural heritage. They will also learn about the application of traditional knowledge and engagement of communities for disaster risk management of cultural heritage. This will help relevant organisations and professionals to develop and implement projects in this field.

The intended audience for this book are Practitioners or site managers of cultural heritage sites and museums. Also, researchers and students studying disaster risk management of cultural heritage. The book will also be of interest to disaster risk management institutions at the urban, regional or national level/cultural heritage management institutions at the urban, regional or national level/city administration, municipalities, urban local bodies and planning departments/educational and research institutions which have specialised programmes in engineering, planning, disaster risk management, and conservation of cultural heritage.

Rohit Jigyasu is a conservation architect and risk management professional from India, currently working at ICCROM as a Project Manager on Urban Heritage, Climate Change and Disaster Risk Management. He is also the Vice President of ICOMOS International Scientific Committee on Risk Preparedness (ICORP). Rohit served as UNESCO Chair holder professor at the Institute for Disaster Mitigation of Urban Cultural Heritage at Ritsumeikan University, Kyoto, Japan, where he was instrumental in developing and teaching the International Training Course on Disaster Risk Management of Cultural Heritage. He was the elected President of ICOMOS India from 2014 to 2018 and president of the ICOMOS International Scientific Committee on Risk Preparedness (ICORP) from 2010 to 2019. Rohit served as the Elected Member of the Executive Committee of ICOMOS since 2011 and was its Vice

President from 2017 to 2020. Before joining ICCROM, Rohit worked with several national and international organisations such as UNESCO, UNISDR, Getty Conservation Institute, and World Bank for consultancy, research, and training on Disaster Risk Management of Cultural Heritage.

Dowon Kim is an associate professor at the Department of Civil and Environmental Engineering, Ritsumeikan University and a UNESCO Co-Chair holder professor at the Institute of Disaster Mitigation for Urban Cultural Heritage, Ritsumeikan University (R-DMUCH). Since 2014, he has been conducting and organising the International Training Course (ITC) on Disaster Risk Management of Cultural Heritage. His current research focuses on community design of disaster risk management in historical districts nationally and internationally. As for international contribution, he works for the International Scientific Committee on Risk Preparedness (ICORP) of ICOMOS. The main objective of his research is sharing the best practices of the urban and local communities' sustainability and cultural/social identity, with the method of analytics, field investigation, and communication tools development, and theorising these practices to transfer to the global context. Moreover, he is interested in disaster mitigation planning to balance heritage conservation, using traditional knowledge and community practices in urban heritage.

Lata Shakya is an associate professor at the Institute of Disaster Mitigation for Urban Cultural Heritage R-DMUCH, Ritsumeikan University. She has been serving as a coordinator of the International Training Course on Disaster Risk Management of Cultural Heritage since 2020. She received her doctoral degree in Urban and Environmental Engineering from Kyoto University and pursued postdoctoral research at the University of Tokyo, Department of Architecture, as a Japan Society for the Promotion of Science postdoctoral fellow and project researcher from 2013 to 2017. Her research focuses on community resilience to disasters and the sustainability of historic cities. She has conducted research in various urban and rural areas of Nepal and Japan. Her primary field area is the historic city of Patan in Nepal, a courtyard-style settlement that originated from Buddhist monasteries. She is currently involved in disaster mitigation planning of different communities in Patan through workshops with the local community. She is the co-author and coordinator of the book 'The Memory of 2015 Nepal Earthquake, the Experience of Local Residents Utilizing Traditional Resources in UNESCO World Heritage Site' (2019) and a co-author of the book 'Rural and Urban Sustainability Governance' (2014). She was awarded 'The Encouragement Prize of AIJ (Architectural Institute of Japan)', the Doctoral Dissertation Award from Association of Urban Housing Sciences in 2014, and the first JUSOKEN Doctoral Dissertation Award in 2016.

Routledge Studies in Hazards, Disaster Risk and Climate Change

Series Editor: Ilan Kelman

Professor of Disasters and Health at the Institute for Risk and Disaster Reduction (IRDR) and the Institute for Global Health (IGH), University College London (UCL)

This series provides a forum for original and vibrant research. It offers contributions from each of these communities as well as innovative titles that examine the links between hazards, disasters and climate change, to bring these schools of thought closer together. This series promotes interdisciplinary scholarly work that is empirically and theoretically informed, with titles reflecting the wealth of research being undertaken in these diverse and exciting fields.

Climate Change and Risk in South and Southeast Asia
Sociopolitical Perspectives
Edited by Devendraraj Madhanagopal, Salim Momtaz

Gender-Based Violence and Layered Disasters
Place, Culture and Survival
Nahid Rezwana & Rachel Pain

The Post-Earthquake City
Disaster and Recovery in Christchurch, New Zealand
Paul Cloke, David Conradson, Eric Pawson & Harvey C. Perkins

Local Adaptation to Climate Change in South India
Challenges and the Future in the Tsunami-hit Coastal Regions
Devendraraj Madhanagopal

Good Practices for Disaster Risk Management of Cultural Heritage
Practices of ITC Participants
Edited by Rohit Jigyasu, Dowon Kim and Lata Shakya

For more information about this series, please visit: www.routledge.com/Routledge-Studies-in-Hazards-Disaster-Risk-and-Climate-Change/book-series/HDC

Good Practices for Disaster Risk Management of Cultural Heritage

Practices of ITC Participants

Edited by Rohit Jigyasu, Dowon Kim and Lata Shakya

Routledge
Taylor & Francis Group
LONDON AND NEW YORK

First published 2023
by Routledge
4 Park Square, Milton Park, Abingdon, Oxon OX14 4RN

and by Routledge
605 Third Avenue, New York, NY 10158

Routledge is an imprint of the Taylor & Francis Group, an informa business

British Library Cataloguing-in-Publication Data
A catalogue record for this book is available from the British Library

Library of Congress Cataloging-in-Publication Data
Names: Jigyasu, Rohit, editor. | Kim, To-wŏn (Associate professor), editor. | Shakya, Lata, editor.
Title: Good practices for disaster risk management of cultural heritage / edited by Rohit Jigyasu, Kim Dowon and Lata Shakya.
Description: London ; New York, NY : Routledge, Taylor & Francis Group, 2022. | Series: Routledge Studies in Hazards, Disaster Risk and Climate Change | Includes bibliographical references and index.
Identifiers: LCCN 2023010007 (print) | LCCN 2023010008 (ebook) | ISBN 9781032411446 (hardback) | ISBN 9781032411453 (paperback) | ISBN 9781003356479 (ebook)
Subjects: LCSH: Cultural property—Protection—Case studies. | Risk management—Case studies.
Classification: LCC CC135 .G658 2022 (print) | LCC CC135 (ebook) | DDC 363.6/90722—dc23/eng/20230411
LC record available at https://lccn.loc.gov/2023010007
LC ebook record available at https://lccn.loc.gov/2023010008

ISBN: 978-1-032-41144-6 (hbk)
ISBN: 978-1-032-41145-3 (pbk)
ISBN: 978-1-003-35647-9 (ebk)

DOI: 10.4324/9781003356479

Typeset in Times New Roman
by Apex CoVantage, LLC

Contents

List of Figures *x*
List of Tables *xii*
List of Contributors *xiii*
Preface *xix*
Acknowledgements *xx*

1 **Introduction** 1
ROHIT JIGYASU, DOWON KIM, AND LATA SHAKYA

PART 1
DRM Frameworks and Risk Assessment 9

2 **Guidance for Preparing Risk Management Plans in New Zealand** 11
VANESSA TANNER

3 **Fire Risk Mitigation Strategies at Urban Heritage Sites** 24
ABDELHAMID SALAH AL-SHARIEF

4 **Ancient Town Walls at Risk: Methods, Technologies, and Tools for Multi-hazard Risk Analysis, Monitoring, and Governance** 37
FRANCESCA GIULIANI

5 **A Cultural Heritage Risk Index – The STORM Project Perspective** 45
MOHAMMAD RAVANKHAH, MARIA JOÃO REVEZ, ROSMARIE DE WIT, ANGELOS CHLIAOUTAKIS, ATHANASIOS V. ARGYRIOU, JOERN BIRKMANN, APOSTOLOS SARRIS, AND MAJA ŽUVELA-ALOISE

PART 2
DRM Plan Implementation – Workshops/Community Engagement/Traditional Knowledge 57

6 **Mapping Risk for Cultural Heritage: A Project on Archaeological Decorative Elements in Mexico** 59
DULCE MARÍA GRIMALDI AND MÓNICA VARGAS

7 **George Town World Heritage City's Disaster Risk Management** 66
MING CHEE ANG

8 **Historic Water Cisterns – An Effective Fire Preventive System** 74
ELENA MAMANI

9 **Participatory methods for DRR in Santo Domingo Tehuantepec, Mexico** 84
DAVID A. TORRES CASTRO

PART 3
DRM Plan Implementation – Stakeholders' Participation/Decision-Making 93

10 **Risk assessment of Humberstone & Santa Laura Saltpetre Works** 95
MARCELA HURTADO

11 **Disaster Risk Management Plan for Punakha Dzong, Bhutan** 107
JUNKO MUKAI

PART 4
DRM Plan Implementation – Capacity Building/Others 117

12 **From Theory to Practice: Insights From the Pathway to Implement DRM Measures for Cultural Heritage Sites** 119
MONIA DEL PINTO

13 Earthquake and Cultural Heritage: A Rescue Project in Central Italy 128
BARBARA CARANZA

14 Future Approach 135
ROHIT JIGYASU, DOWON KIM, AND LATA SHAKYA

Index *139*

Figures

1.1	Disaster Risk Management Cycle for Cultural Heritage Sites	3
2.1	Christ Church Cathedral damaged during the 2010–2011 earthquakes is a symbol of the struggle to retain heritage buildings	12
2.2	Risk management process based on AS/NZS ISO 31000:2009	15
2.3	Disaster risk cycle	17
2.4	Risk evaluation matrix	19
3.1	New buildings around historical buildings	26
3.2	Proposed fire risk assessment methodology for fire mitigation for the selected site	27
3.3	Demolition of old buildings, Atfit Hosh el-Nimr	28
3.4	Mitigation measures for fire primary hazard based on the risk assessment	30
4.1	Cities still preserving ancient town walls in Tuscany (Italy)	38
4.2	Emblematic examples of failures of town walls	41
4.3	Preliminary results of the analysis on the town walls of Pisa	42
5.1	The concept of risk index and risk map for the pilot sites of the project	47
5.2	Developing an earthquake risk index (on the left) and an earthquake risk map (on the right) for Tróia	51
5.3	Situational awareness Web-GIS service of the STORM platform for the 'Historical Centre of Rethymno' pilot site	53
6.1	Collapsed shelter over the wall painting and stucco reliefs and pavements due to Hurricane Grace, at the Archaeological Site of Tajín, Mexico	60
6.2	Risk and priority map for the case study of the decorative elements at the Archaeological Site of Cholula, Mexico	63
8.1	Map of identified water cisterns within the historical centre	78
8.2	The hydrants	79
8.3	The final test	81

9.1 Disaster Imagination Game tool used to identify threats, vulnerabilities, strengths, and mitigation strategies during the participatory workshop with local stakeholders 87
9.2 Vulnerability and risk maps generated through participatory tools 89
10.1 World Heritage Site Humberstone and Santa Laura Saltpetre Works, Pozo Almonte, Chile 96
10.2 Methodology for the development of the disaster risk management plan 97
10.3 Risk assessment for some of the buildings in both components of the site 100
10.4 Proposal of activities for disaster risk reduction for Humberstone and Santa Laura Saltpetre Works 102
10.5 Articulation proposal between different stakeholders for disaster risk reduction in the site 103
11.1 Punakha Dzong 108
11.2 Wangduephodrang Dzong in fire in 2012 114
12.1 Summary of steps undertaken between December 2019 and August 2020 123
12.2 The stakeholders outlined for the reviewed project 124
13.1 The CHIEF ETS logo 129
13.2 San Salvatore a Campi – the volunteers of CHIEF ETS during the rescue phases 131
13.3 San Salvatore a Campi – detail of the operations of consolidation of the paint film 132

Tables

12.1 Summary of hazard and vulnerability assessment in the MuNDA 120

Contributors

Abdelhamid Salah al-Sharief is the Chairman of Egyptian Heritage Rescue Foundation, Risk Assessment Unit manager at the Ministry of Tourism and Antiquities, and Consultant and resource person on Crises Risk Management in the National Training Academy. He is also a Consultant and project co-manager with ICCROM. He holds a Master's degree in Heritage Conservation and Site Management from Helwan University and he is a PhD candidate at Brandenburg University in the Heritage and Museum Studies Program. He has also pursued Bachelor of Arts from Faculty of Archaeology Cairo University and Wood Conservation Technology at the Norwegian University of Science and Technology. He has also attended the International Training Course on Disaster Risk Management of Cultural Heritage, by Institute of Disaster Mitigation for Urban Cultural Heritage, Ritsumeikan University.

Angelos Chliaoutakis received his PhD in Electronic & Computer Engineering in 2020 at Technical University of Crete (TUC), Greece. Since 2010 he is a research assistant at the Laboratory of Geophysical - Satellite Remote Sensing and Archaeo-environment (GeoSat ReSeArch Lab) of the Institute for Mediterranean Studies of the Foundation for Research and Technology, Hellas (IMS-FORTH). He is involved in various research projects related to the full-stack development of Geographical Information Systems (GIS), web-based GIS applications and Geoinformatics in the cultural, archaeological and environmental domain.

Apostolos Sarris is Professor of the "Sylvia Ioannou" Chair on Digital Humanities at the Archaeological Research Unit, Department of History and Archaeology, University of Cyprus and Director of the DigHumanities GeoInformatics Lab. His research is focused on Geophysical Prospection, GIS spatial modelling and satellite remote sensing (260 projects in Greece, U.S.A., Cyprus, Hungary, Albania, Germany, Italy, Turkey, Serbia, Israel, and Egypt). He is a corresponding member of the German Archaeological Institute and member of ISAP, UISPP, CAA, EAA and SAA.

Athanasios V. Argyriou is a Researcher C in Eratosthenes Centre of Excellence (ECoE), Cyprus. His research focuses on geoinformatics applications (GIS / Remote Sensing / Geophysics) on natural hazard assessment, geomorphometrics, land use/cover changes, urban sprawl, change detection analysis, preservation of cultural heritage, historical mapping, digital humanities and archaeological landscape. He has participated in various international projects with his main interests focusing on: (a) earth and environmental sciences; (b) providing geo-information and cartographic solutions in environmental sciences, disaster risk reduction from natural hazards; (c) spatial analysis on digital humanities, historical mapping and archaeological issues; and (d) application of geophysical prospection surveys.

Barbara Caranza is a restorer of stone materials and derivatives, decorated surfaces of architecture, and painted artefacts on wooden and textile support. She is a professor for higher education schools and universities and for the Civil Protection system in matters concerning the protection of cultural heritage in crisis areas, methodologies of stabilisation, and disaster risk management plans for Cultural Heritage. She serves as the Officer of selected reserve of the Italian Army for Cultural Property Protection and numerous other emergency planning and management interventions in reference to cultural heritage in Italian and foreign operational theatres through ministerial assignments. She is the president and founder of Cultural Heritage International Emergency Force.

David A. Torres Castro is an object conservator specialised in disaster risk management for cultural heritage. He has an MSc in Risk, Disaster and Resilience from the Institute for Risk and Disaster Reduction, University College London (UK), and a BA from the National School of Conservation (ENCRyM, Mexico). He is passionate about heritage and disaster management. He has also studied in the Institute of Disaster Mitigation for Urban Cultural Heritage, in Ritsumeikan University (Japan) as part of a UNESCO training programme. His academic training allowed him to experience on-the-ground emergency responses, conduct disaster risk research, and explore cultural heritage recovery processes in different contexts.

Dulce María Grimaldi is a senior conservator who has served since 1994 for the National Institute of Anthropology and History of Mexico (INAH). Over 20 years, she has focused on the conservation of decorative elements at archaeological sites, mainly in the central part of Mexico. As well, she has collaborated with international conservation projects, such as the Conservation Project of TT39 at Luxor, Egypt. As part of the conservation projects, mostly on wall painting, stone, and stucco reliefs, she has promoted the disaster risk management as part of the work for the preservation of cultural heritage in this region.

Elena Mamani is an award-winning conservation architect with a life-long dedication to the culture of Albania. With the knowledge and experience that she has gained internationally, she has executed over a dozen of conservation and emergency preparedness and response projects. Elena graduated in Architecture Engineering from the National Technical University of Athens in 2004. Her engagement in the field of cultural heritage began in 2008 with the first project of CHwB in Albania. She is currently responsible for the implementation of full restoration projects and training of architects in the field of restoration/conservation and risk preparedness and coordination and overall management of educational programmes for young professionals in the field of conservation and vocational education of craftspeople.

Francesca Giuliani is a Chartered Civil Engineer with a master's degree in Architectural Engineering at the University of Pisa (2016) and a PhD in Structural Engineering at the University of Pisa (2020) awarded with the additional certification of Doctor Europaeus. Currently, she is a post-doctoral researcher at the Department of Civil Engineering, University of Pisa, where she is developing interdisciplinary research in Earthquake Engineering and Disaster Risk Reduction. In the last few years, she has cooperated with diverse national and international research institutes, developing interdisciplinary projects focused on the safety and safeguard of urban cultural heritage.

Joern Birkmann, Prof. Dr.-Ing. habil., is director at the Institute of Spatial and Regional Planning of the University of Stuttgart. In addition to issues of planning and spatial governance, he has particularly worked in the area of vulnerability and risk research in the context of natural disasters and climate change in recent years. Publications in Nature and Nature Climate Change underscore his international expertise and visibility. His research has also informed and contributed to international reports, such as UN and World Bank reports. He was the coordinating lead author for the sixth assessment report of the IPCC.

Junko Mukai is Assistant Professor at the Graduate School of Design and Architecture, Nagoya City University, and working on a project to improve the seismic resilience of traditional structures in Bhutan. From 2000 to 2016, worked at the Department of Culture (DoC), Ministry of Home and Cultural Affairs, Government of Bhutan as a heritage architect to establish the legal and institutional framework for heritage protection and develop DoC's technical capacity through various project implementations.

Later worked as a consultant for the DoC, World Bank, and UNESCO on heritage-related projects. She is also the member of ICOMOS Japan, Okinawa Society of Architects and Aichi Heritage Conference.

Maja Žuvela-Aloise works at the GeoSphere Austria - Federal Institute for Geology, Geophysics, Climatology and Meteorology, former ZAMG, since 2010. She holds a diploma degree (DI) in Physics and Geophysics from the University of Zagreb, Croatia and a PhD from the University of Kiel, Germany. Her expertise lies in regional and urban climate modelling for investigation of climate change and climate adaptation. Her research is focused on Urban Heat Island mitigation using nature-based solutions and application of climate information in sustainable urban development and spatial planning.

Marcela Hurtado is an associate professor of the Department of Architecture at the Federico Santa Maria Technical University, Valparaiso, Chile. She is an architect from the Valparaiso University and a specialist in conservation and restoration from the University of Chile. She received her PhD in Art History and Latin-American architecture from the Pablo de Olavide University, Sevilla, Spain. Her research interests include risk management in cultural heritage, construction history, and architecture history. She has worked in several projects related to the management of Chilean cultural heritage. She is the director of a postgraduate programme in sustainable rehabilitation in architecture at the Federico Santa Maria Technical University, president of the Chilean Committee of the International Council on Monuments and Sites (ICOMOS), and vice-president of the International Scientific Committee on the Analysis and Restoration of Structures of Architectural Heritage (ISCARSAH).

Maria João Revez is a conservator-restorer holding a PhD in Heritage Conservation and Restoration from the NOVA School of Science and Technology. Currently, she works for Nova Conservação, a Portuguese SME dedicated to built heritage conservation, as the main responsible for the company's R&D efforts, including the participation in Project STORM, an H2020-funded project on the managing of disaster risks to archaeological heritage sites. Maria João is also an invited assistant professor at the Polytechnic Institute of Tomar, where she lectures on Philosophy of Conservation and Heritage Risk Management in the master's programme in Conservation and Restoration.

Ming Chee Ang is the General Manager of George Town World Heritage Incorporated. She is also an accredited facilitator of UNESCO Global Network of Facilitators on Intangible Cultural Heritage, and an alumnus cum resource person of UNESCO Chair Programme on Cultural Heritage and Risk Management International Training Course on Disaster Risk Management of Cultural Heritage. Born and raised in the inner city of George Town, Ang carries her duties as a World Heritage Site Manager with much passion and fervour. Specialised in resource mobilisation, policymaking, project management, and risk assessment, Ang has incorporated building conservation, disaster risk reduction, and intangible cultural heritage safeguarding to create a better and more resilient heritage city in George Town.

Mohammad Ravankhah (Ravan) is a postdoctoral research fellow in the Institute of Spatial and Regional Planning at the University of Stuttgart. He did his PhD in disaster risk assessment methodology for World Heritage Sites. His expertise falls in the areas of disaster risk management and climate change adaptation, in particular developing vulnerability and risk assessment methodologies and tools. He is particularly interested in interdisciplinary research to enhance disaster resilience and climate adaptation in urban heritage, urban planning, and regional development.

Dr. Monia Del Pinto is a Doctoral Prize Fellow in the School of Architecture, Building and Civil Engineering at Loughborough University. She owns a PhD in urban planning and disaster risk reduction and her research focuses on urban morphology and urban vulnerability in heritage urban areas. Monia has been involved in post-earthquake reconstruction in historical contexts in Central Italy and is active in the field of disaster risk management for cultural heritage sites.

Mónica Vargas is a conservator at the National Institute of Anthropology and History of Mexico (INAH).

Rosmarie de Wit obtained a degree (MSc) in Physics (program stream Meteorology, Physical Oceanography and Climate) from Utrecht University, the Netherlands. During her PhD at the Norwegian University of Science and Technology in Trondheim, she specialized in atmospheric observations. After a postdoc at NASA Goddard Space Flight Center gaining experience in working with large observational data sets, she joined the urban modelling group at ZAMG in 2016. Her work is focused on climate change and climate change adaptation, as well as ways to communicate these results. Since 2020 she is in charge for the public relations of the ZAMG, now GeoSphere Austria.

Vanessa Tanner is currently the Manager of Archaeology for Heritage New Zealand Pouhere Taonga and is based in Wellington, New Zealand. Previously, Vanessa worked for Local Government in New Zealand, in various heritage management and archaeology roles; projects include managing funding for seismically strengthening earthquake-prone buildings and developing a condition/threat/response monitoring method for managing archaeological sites facing climate change. Following attendance at the 2016 UNESCO Chair International Training Course on Disaster Risk Management of Cultural Heritage (ITC), Vanessa became a member of the Australia New Zealand ICOMOS Joint Scientific Committee on Risk Preparedness (JSC-ANZCORP).

Preface

Earthquakes, tsunamis, and other such calamities are increasingly shaking the foundation of human life, culture, and heritage. The world recently witnessed Turkey and Syria shattered to rubbles in an earthquake. While still recovering from the earthquakes of Central Mexico in 2017, Myanmar and Italy in 2016, and Nepal in 2015. The devastating tsunami in the north-east of Japan in 2011 and fire in the Notre-Dame de Paris in 2019, National Museum of Brazil in 2018, and Lijiang, China in 2013 have depleted the world's cultural heritage. Apart from natural calamities, the ongoing war in Ukraine and the COVID-19 pandemic have caused enormous loss of life, property, and cultural heritage, both in its tangible and intangible as well as movable and immovable manifestations. These have highlighted the fact that cultural heritage, including historic buildings, archaeological sites, historic cities, cultural landscapes, and museum collections are highly vulnerable to disasters caused by natural and human-induced hazards. Furthermore, climate change is predominantly causing an increase in the frequency and intensity of hydro-meteorological hazards such as floods and typhoons/cyclones.

Hence, it is important to adopt proactive measures reducing the risks to cultural heritage from these catastrophic events through adequate mitigation and preparedness measures. During emergency phase, the challenge is how to assess damage and stabilise built heritage properties, which are at risk of demolition and salvage movable heritage fragments and collections and assess their damage. The long-term challenge during recovery phase is repairing, retrofitting, and reconstructing the tangible as well as intangible heritage values while reducing vulnerabilities. In the light of these challenges, comprehensive disaster risk management is essential for the protection of cultural heritage from disasters. Since 2006, Institute of Disaster Mitigation for Urban Cultural Heritage, Ritsumeikan University (R-DMUCH) has been organising an International Training Course with the aim of building capacities of heritage and disaster risk management professionals.

This book highlights various aspects such as integrated risk assessment, community participation, use of traditional knowledge, appropriate decision-making process for implementing DRM plans, sustainable mitigation, and preparedness strategies through best practices of professionals and frontline workers trained through ITC, in disaster risk management of cultural heritage.

Acknowledgements

We would like to thank all the former participants of the International Training Course, who have contributed to this publication based on their experience in developing and implementing various activities for disaster risk management of cultural heritage. We are also extremely grateful to our resource people, who reviewed the chapters and provided valuable suggestions to enhance this publication. These include Joseph King, Lee Bosher, Ksenia Chmutina, Wesley Cheek, Ming Chee Ang, Takeyuki Okubo, Barbara Minguez Garcia, Xavier Romão, and Chris Marrion. Lastly, we express our heartfelt thanks to the staff of R-DMUCH, who have helped us immensely through all the process.

1 Introduction

Rohit Jigyasu, Dowon Kim, and Lata Shakya

1. The International Training Course on Disaster Risk Management of Cultural Heritage

History

The International Training Course (ITC) on Disaster Risk Management of Cultural Heritage is a follow-up of the recommendations adopted at the Special Thematic Session on Risk Management for Cultural Heritage held at the UN World Conference on Disaster Reduction (UN-WCDR) in January 2005 in Kobe, Hyogo, Japan. One of the recommendations advocated the need for the academic community to develop scientific research, education, and training programmes incorporating cultural heritage in both its tangible and intangible manifestations into disaster risk management. The World Heritage Committee reiterated the importance of strengthening knowledge, innovation, and education to build a culture of disaster prevention at World Heritage (WH at its 30th session (Vilnius, Lithuania, July 2006)).

Furthermore, the 'Declaration' adopted at the International Disaster Reduction Conference (IDRC) of Davos (August 2006) confirmed that 'concern for heritage, both tangible and intangible, should be incorporated into disaster risk reduction strategies and plans, which are strengthened through attention to cultural attributes and traditional knowledge'. The Sendai Framework on Disaster Risk Reduction recently adopted at the World Conference on Disaster Risk Reduction in Sendai, Japan, further highlighted the importance of protecting cultural heritage from disasters.

The Institute of Disaster Mitigation for Urban Cultural Heritage at Ritsumeikan University (R-DMUCH) has been responding to these recommendations by the international community by organising international research, training, and information network in the field of cultural heritage risk management and disaster mitigation.

These training courses have welcomed *180 participants from 72 countries* across the world.

DOI: 10.4324/9781003356479-1

Objectives of ITC

ITC aims to promote an intensive educational programme, scientific networking, and research on disaster risk management of cultural heritage. It aims to create sufficient measures for cultural heritage sites to reduce risks to both movable and immovable, as well as tangible and intangible cultural heritage. These disasters are caused by natural hazards such as earthquakes, tsunamis, floods, typhoons, landslides, and forest fires and by human-induced hazards like arson, vandalism, terrorism, and conflicts, including biological hazards.

The main objective of the course is to provide theoretical and practical knowledge on various aspects of disaster risk management of cultural heritage. The course is an amalgamation of lectures, site visits, workshops, discussions, team projects, and individual/group presentations. Participants are expected to pursue the entire course actively. The course promotes collaborations and network-building among scholars and professionals in cultural heritage protection. This course is scientifically supported by the International Centre for the Study of the Preservation and Restoration of Cultural Property (ICCROM).

Based on the knowledge obtained from lectures, site visits, and exercises through interactive workshops, the training course also encourages planning in disaster risk management of cultural heritage. As part of the course, each participant is required to formulate an outline of a DRM plan for a case study site or museum from their home country in line with the country's respective social, economic, and institutional context. As a prerequisite, the selected participants are instructed to collect relevant data/information related to the cultural heritage, hazard characteristics, and local context before coming to Japan.

A Key Theme of DRM of Cultural Heritage in ITC

Disaster risk management is a cyclical process with three basic stages: before, during, and after a disaster. Before a disaster, the main activities include risk assessment, prevention, mitigation methods, and warning systems for specific hazards. Planning for emergency evacuation and response procedures are all activities that should be undertaken in advance for responding during a disaster. Activities initiated after the disaster include damage assessment, treatment of damaged components of the heritage property through interventions such as repairs, restoration and retrofitting, and recovery or rehabilitation activities. At this stage, the effectiveness of the previous stages can also be evaluated for proactive prevention of any subsequent event. An integrated approach to disaster risk management of cultural heritage considers the possibilities of multiple hazards that may occur in parallel or as follow-up due to interactions between various natural and human-induced causes. It stresses the importance of community participation and ensures regular maintenance and management procedures for the site. It also highlights the importance

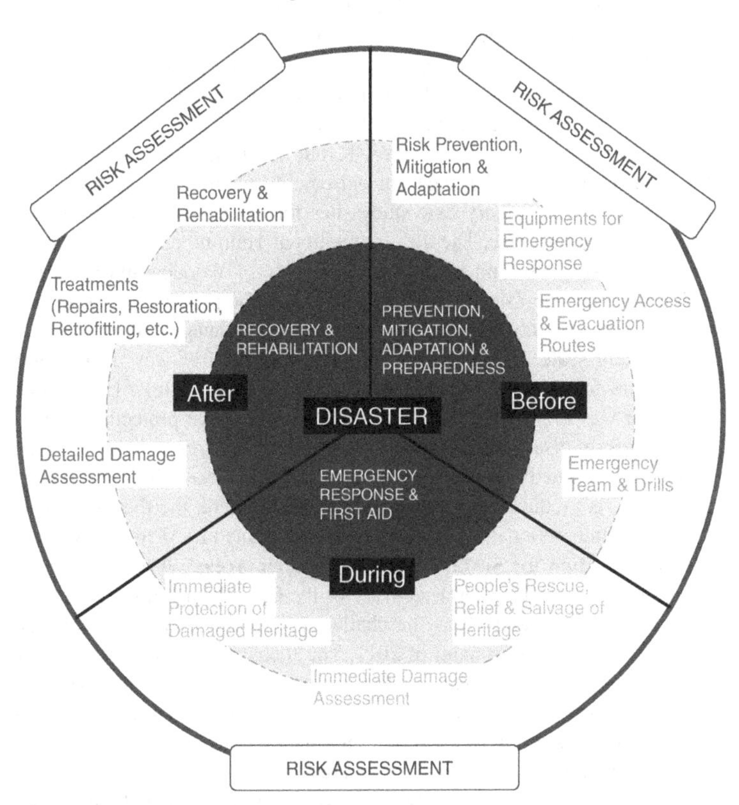

Figure 1.1 Disaster Risk Management Cycle for Cultural Heritage Sites.

Source: Authors (Rohit Jigyasu and Dowon Kim)

of having a proactive role in cultural heritage in reducing disaster risks. As outlined earlier, the disaster risk management framework forms the basis of planning the course.

Themes of Previous International Training Courses (2006–2021)

The International Training Course has evolved significantly since its first session in 2006, consisting of eight participants from four countries where major disasters had struck. In 2007, the course was modified to closely interlink lectures, site visits, and workshops based on the first-year experience.

The second year also focused on participants developing their own disaster risk management plans, which enabled interactive training sessions and the knowledge exchange between participants. In its third year, the training course was revised significantly, and the primary focus shifted towards making it field-based. Most workshops conducted were based on site visits to the World Heritage Sites in Kyoto, Japan. In the fourth session held in 2009, the training course introduced a specific thematic focus titled *Earthquake Risk Management of Historic Urban Areas*. Kyoto and Kathmandu, two historic cities with rich cultural heritage but extremely vulnerable to earthquakes, were chosen as the primary case study sites for undertaking field exercises during the training course. The fifth International Training course focused on 'Emergency Response and Long-Term Recovery of Wooden and Composite Cultural Heritage from Earthquake and Fire'. The pedagogical strategies emphasised community engagement in disaster risk management planning at the settlement scale. The theme of the sixth year in 2011 was **'Integrated Approaches for Disaster Risk Mitigation of Historic Cities'**. The goal of this course was to introduce participants to the proactive protection of historic cities from disasters and develop mitigation measures to be undertaken through an integrated approach. Given the growing concerns for mainstreaming disaster risk reduction into sustainable development, the thematic focus of the 2012 International Training Course was framed as **'From Recovery to Risk Reduction for Sustainability of Historic Areas'**. It was built upon the lessons from the long-term recovery of the Great Hanshin Awaji (Kobe) Earthquake of 1995 as well as the challenges presented by the Great East Japan Earthquake and Tsunami of 2011. The focus was on the post-disaster recovery of cultural heritage, extending beyond restoration to the revival of tangible and intangible aspects of heritage. It emphasised effective engagement with various stakeholders at the city, national and international levels for protecting cultural heritage in urban areas during catastrophic situations. In 2013, the course focused on **'Reducing Disaster Risks to Historic Urban Areas and Their Territorial Settings Through Mitigation'**, and in the 2014 course, the spotlight was on **'Protecting Living Cultural Heritage From Disaster Risks Due to Fire'**. Policies and planning measures for reducing fire risks to cultural heritage, especially in the rapidly urbanising context of developing countries, were highlighted. Also, special techniques for fire prevention and mitigation, emergency response, and interventions for long-term restoration and rehabilitation of cultural heritage following disaster were discussed. The main theme of the 2015 course was **'The Protection of Cultural Heritage From Earthquakes and Floods, and Other Associated Hazards'** and the 2016 course emphasised **'Protecting Cultural Heritage From Risks of Natural Disasters Including Those Induced by Climate Change'**, focusing on climate change-caused hazards. The theme for 2017, 2018, and 2019 was **'The Integrated Approach for Movable and Immovable Heritage for Disaster Risk Management of Heritage Sites as well as Museums'**, focusing on the concept of movable and immovable heritage before, during, and after a disaster situation.

In 2020, the training course was cancelled due to the COVID-19 pandemic. Alternative programmes were organised, such as the webinar series **'Capacity Building for Disaster Risk Management of Cultural Heritage: Challenges and Opportunities in Post-COVID Times'** held on June 27 and July 04, 2020 and a workshop on **'Good Practices for Disaster Risk Management of Cultural Heritage'** organised on October 8–10, 2020. During this workshop, good practices were identified, which led to this publication.

In 2021, another online ITC was organised focusing on 'Disaster Risk Management of Cultural Heritage: Learning From Japanese Experience'. The recent (2022) training course is focused on Japanese experiences in each of the phases of the DRM cycle: risk assessment, mitigation and preparedness, emergency response, recovery, and policies. To provide an effective online course, three sessions – a preparatory session (lecture videos and site visit videos), an interactive live session (workshops, group work, and group discussions), and a post-interactive session (mentoring and feedback from resource persons for the case study projects) – were prepared for each phase. While Zoom was used as a virtual platform for the interactive live sessions, 'Slack' was useful for formal and informal communication among the resource persons and participants. All the teaching resources as well as participants' case studies were posted on google drive before the interactive live sessions for easy access. The COVID-19 pandemic posed some significant challenges for conducting fieldwork. However, alternative video tours of local sites were created with the cooperation of the heritage site owners and the local community in the city of Kyoto. In 2022, we successfully organised an online course on the theme 'Traditional knowledge for DRM of Cultural Heritage'. There were 15 participants from 14 countries, and they developed a disaster risk management plan for their case study sites. The course outcomes in the form of DRM plans are published as proceedings each year in March.

2. Why Is This Book Necessary?

The International Training Course (ITC) has been an annual course since 2006, and among the 180 participants, many are working on the frontlines of DRM of cultural heritage. There are several books and articles on the DRM plans of CH, but they lack concrete examples from practice. This book discusses the practical experience of experts working for the DRM of CH, along with the former ITC workshop participants in 2020. This would be beneficial in developing and implementing DRM plans.

The **'Good Practices for Disaster Risk Management of Cultural Heritage'** workshop held on October 8–10, 2020 aimed to showcase various projects on disaster risk management of cultural heritage undertaken by the former ITC participants organised by R-DMUCH and ICCROM since 2006.

Additionally, it aimed to review the activities of ITC since 2006 and work towards building a stronger network among the ITC lecturers and former participants. The call for applications was launched for the former ITC participants who had participated in the ITC since 2006. A total of 27 applications were received and 7 participants were selected to present their projects during the workshop. After deliberation by the jury members, two were selected for the 'Best Practice Award' and one presenter was selected for the 'Exemplary Practice Award'.

Besides these seven participants, the organising committee selected five additional good practices wherein participants were asked to submit their manuscripts on good practices in disaster risk management of cultural heritage. This book is a collection of these submitted manuscripts on good practices from 16 countries. The book has 14 chapters divided into various themes including formulation of policies, risk assessment, and implementation of various DRM strategies for respective cultural heritage sites and institutions.

3. What Will You Find in This Book?

Among various topics of Disaster Risk Management, this book covers case studies that discuss methodologies, tools, and strategies for assessment, mitigation, preparedness, response, and recovery, both in terms of planning and implementation in various contexts. The book is divided into 14 chapters. The chapters are distributed among four broad topics:

a) PART 1 DRM frameworks and risk assessment
b) PART 2 DRM plan implementation – workshops/community engagement/traditional knowledge
c) PART 3 DRM plan implementation – stakeholders' participation/decision-making
d) PART 4 DRM plan implementation – capacity building/others

Tanner (Chapter 2) presents the process of DRM planning, and Abdelhamid (Chapter 3), Giuliani (Chapter 4), and Ravankhah (Chapter 5) focus on the risk assessment process. The brief legislative background and requirements of the DRM plan for heritage sites are described in Chapter 2, whereas Chapter 3 is about the fire risk assessment and mitigation strategies of Egypt. Chapters 4 and 5 discuss methods/technologies for risk analysis and risk assessment.

Three topics are covered regarding the implementation process for DRM plans for cultural heritage sites and institutions. Community participation is one of the essential parts of the DRM implementation process. Chapters 6 and 9 describe a tool for risk and vulnerability assessment used with community engagement through a workshop using maps. Chapter 7 describes community-based activities for implementing the DRM plan, and Chapter 8 showcases a success story of using traditional knowledge for implementing a DRM strategy using traditional water cisterns for fire prevention measures. It also highlights the involvement of the community in decision-making.

Decision-making by relevant stakeholders is another essential part of the DRM plan implementation phase. Chapter 10 describes in detail the decision-making process with related stakeholders. Chapter 11 demonstrates the decision-making process within the religious organisation, which is the key custodian of the site and the government policies during the implementation of the very first project of DRM of the heritage site in Bhutan.

Furthermore, the capacity building of professionals from relevant authorities and the community is also essential for implementing the DRM plan. Chapter 12 describes the gap between the formulation of the DRM plan and its implementation on the ground. It summarises the lessons learnt from the organisation of capacity-building projects on DRM of CH. Chapter 13 describes the importance of the capacity building of volunteers for the salvage of cultural heritage.

Lastly, Chapter 14 provides an overview of the key issues and lessons based on the case studies and discusses future approaches.

This publication will contribute towards building the capacity of professionals from cultural heritage and disaster risk management to build the resilience of our cultural heritage sites and institutions against disasters.

Part 1

DRM Frameworks and Risk Assessment

2 Guidance for Preparing Risk Management Plans in New Zealand

Vanessa Tanner

Introduction

New Zealand has a habitation history of 800 years, beginning with Māori, who came from the Pacific in approximately 1300 AD and occupied the country, and whose heritage is embodied in this land. Following European exploration in the 17th and 18th centuries, the country was colonised by Europeans. A treaty was signed between Māori and the British Crown in 1840 to protect Māori culture and represent the interests of all New Zealanders. However, the treaty was not honoured, and this has shaped the country's recent history. In 1975, the British Crown agreed to address this through legislation, and this process continues to shape the country's present. The cultural heritage of New Zealand includes the tangible and intangible heritage of Māori, its subsequent settlers, and the interaction of these people over time.

On September 4, 2010, a 7.1 magnitude earthquake occurred near the city of Christchurch in New Zealand.[1] The aftershocks continued, and on February 22, 2011, the city was severely damaged again by a 6.3 magnitude earthquake, which killed 185 people and injured several thousand.

The buildings and roads across the Christchurch region, weakened by the September 2010 earthquake and its aftershocks, were severely damaged and destroyed in the February 2011 earthquake. Many buildings in the central business district and tens of thousands of dwellings were damaged and people were made homeless.[2]

The Christchurch earthquakes led to a significant loss of the city's heritage, where approximately 50% of the listed built heritage was destroyed.[3] There were a number of reasons for this:

- Heritage was not part of emergency planning or Civil Defence operating procedures,

DOI: 10.4324/9781003356479- 3

Figure 2.1 Christ Church Cathedral damaged during the 2010–2011 earthquakes is a symbol of the struggle to retain heritage buildings. The decision to reinstate the Cathedral was made in 2017.

Source: Photo by Dave Margetts HNZPT November 2, 2017

- Legislation protecting heritage was suspended, and
- The Canterbury Earthquake Recovery Agency which was responsible for the recovery did not consider heritage when demolition orders were issued.[4]

Following the Christchurch earthquakes, and likely due to international developments in disaster risk management and resilience,[5] changes in legislation were enacted to recognise heritage. For example, heritage has been included in New Zealand's Civil Defence and Emergency Management Act 2016. Christ Church Cathedral (see Figure 2.1), damaged during the 2010 and 2011 earthquakes, symbolises the struggle to retain heritage buildings.

Legislative Context

New Zealand is a constitutional monarchy with a parliamentary government system. New Zealand's cultural heritage places are provided for in various statutes, primarily the Heritage New Zealand Pouhere Taonga Act 2014 (HNZPT

Act) and the Resource Management Act 1991. The most relevant legislative provision for this chapter was the addition of a new category of historic place in New Zealand, the National Historic Landmarks/Ngā Manawhenua o Aotearoa me ōna Kōrero Tūturu, to the HNZPT Act. This inclusion aimed to promote the conservation and protection of the most significant heritage places from natural disasters.[6] The reference to natural disasters in the legislation is limited in terms of the risk management plans for heritage places. Therefore, HNZPT's guidance on risk management plans recognises that disasters result from both natural and human-induced hazards. The HNZPT Act requires that an owner has an appropriate risk management plan approved by HNZPT before it can be included on the National Historic Landmarks List.[7]

Working Group Established

In 2018, an HNZPT internal working group was formed to determine what a risk management plan needed to contain to meet the requirement of the HNZPT Act. This working group was made up of nine HNZPT staff from across the organisation and country, with a range of heritage management skills. Some had experience managing places affected by disasters like the Christchurch earthquakes. Others had been involved in preparing heritage properties owned by HNZPT for disasters.

In developing the guideline, the working group drew upon international guidance for managing risk to heritage places. The skills and resources obtained during the UNESCO Chair Programme on Cultural Heritage and Risk Management, International Training Course (ITC) on Disaster Risk Management of Cultural Heritage held at Ritsumeikan University in 2016, were also adopted.

In addition, the working group collaborated on the development of a risk management plan for the first place to be entered in the National Historic Landmarks List, which was Te Pitowhenua/Waitangi Treaty Grounds. The Te Pitowhenua/Waitangi Treaty Grounds plan helped the working group identify and define eight steps to creating a risk management plan, which provided a model for the development of the subsequent guideline.

The draft guideline was reviewed by members of the Australia New Zealand Joint Scientific Committee on Risk Preparedness (JSC-ANZCORP).

Case Study: National Historic Landmark Te Pitowhenua/Waitangi Treaty Grounds

Te Pitowhenua is the most symbolically important place in New Zealand. Te Tiriti o Waitangi/The Treaty of Waitangi, New Zealand's founding document, was signed here in 1840. Hence, it is fundamental to New Zealand's cultural identity and its origins as a modern bicultural nation. In pre-treaty times, this was a place where rangatira (chiefs) gathered to discuss matters of common interest and where nationally important constitutional events took place. Te Pitowhenua is still a pivotal place to engage with ideas about nationhood and national identity. For many Māori and Pākehā, Te Pitowhenua is a living, breathing entity; linked with the aims and aspirations of Te Tiriti and the birthplace of a nation. Te Pitowhenua is central to New Zealand's past, present, and future. It has played a crucial role in social, constitutional, and cultural history and incorporates physical elements of significant value. It also has spiritual importance and ancestral associations with many significant tūpuna for Māori.[8]

The disaster risk management plan for Te Pitowhenua/Waitangi Treaty Grounds was prepared in collaboration with the Waitangi National Trust, which administers the property. Te Pitowhenua/Waitangi Treaty Grounds had already prepared documents like an emergency response plan, these provided a starting point for the risk management plan. The format and content of the Te Pitowhenua/Waitangi Treaty Grounds plan formed the model for the guideline that the working group subsequently developed.

In preparing the Te Pitowhenua/Waitangi Treaty Grounds plan, HNZPT utilised the New Zealand Civil Defence and Emergency Management (CDEM) guidance[9] and the New Zealand Standard AS/NZS ISO 31000:2009[10] risk management planning approach. In addition, the Civil Defence Northland Emergency Management Group Community Response Plan[11] was used to identify local hazards and response plans. These documents, especially the New Zealand Standard proved to be a useful framework which was also utilised internationally in developing disaster risk management plans for cultural heritage places.[12] The New Zealand Standard outlines a process to develop a risk management plan, this is followed in HNZPT's guideline with a cultural heritage perspective added (see Figure 2.2). These documents highlighted the importance of using a common language of risk management and New Zealand's existing disaster risk and emergency management framework.

The Te Pitowhenua/Waitangi Treaty Grounds experience reinforced various insights for the working group. Firstly, HNZPT needed to create a guidance document that would assist owners and managers of heritage places in understanding the value and process of preparing a risk management plan. Secondly, guidance to promote early stakeholder and community engagement is pertinent, so the guideline needed to address that early in the document and dedicated a section to it. The working group learnt that the guideline had to present a step-by-step process resulting in a set of practical actions for heritage owners and managers to undertake.

From the feedback HNZPT has received on the Te Pitowhenua/Waitangi Treaty Grounds plan, it is understood that the plan has assisted the Waitangi National Trust in engaging with their local Civil Defence. Along with the National Historic Landmarks status, it has assisted the Waitangi National

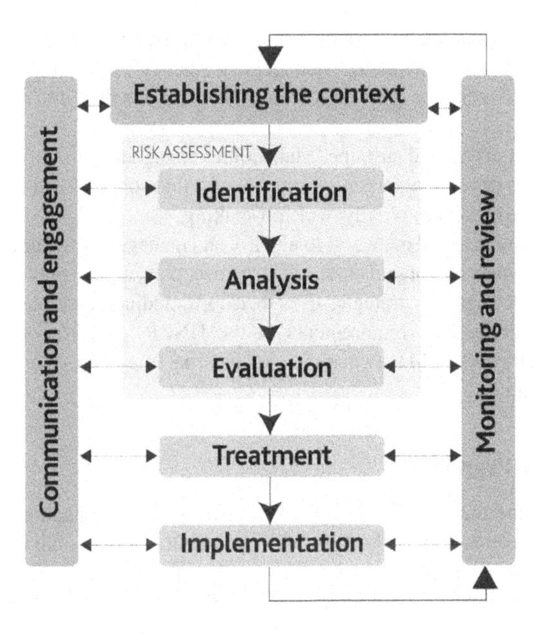

Figure 2.2 Risk management process based on AS/NZS ISO 31000:2009.

Source: Heritage New Zealand Pouhere Taonga, 2022

Trust with a funding application. Since the plan included a section on pandemics, the Waitangi National Trust was aware that the COVID-19 pandemic would impact their site.

Guideline Preparation

While developing the Te Pitowhenua/Waitangi Treaty Grounds risk management plan, the working group began drafting HNZPT's risk management guideline for owners and managers of heritage places.

Initially, the working group focused only on addressing HNZPT's National Heritage Landmarks legislative requirements. However, the scope was expanded to include any heritage place in New Zealand.

The objectives of the working group were to create a guideline that:

- Proposes a step-by-step guide with clear instructions to owners and managers on preparing risk management plans.

- Aligns with New Zealand's existing risk management framework by using the 4 Rs methodology for risk management planning, which is recognised by the National Emergency Management Agency.[13]
- Applies to a diverse range of heritage places, ownership, and management models.
- Promotes active and sustained stakeholder engagement.
- Promotes developing organisational culture that supports and incorporates risk management into daily decision-making.
- Enables owners and managers to identify and manage risk to heritage places.
- Helps owners and organisations to prioritise, plan, and resource risk reduction and mitigation strategies over the short, medium, and long term.
- Meets the specific requirements of the HNZPT Act National Historic Landmarks/Ngā Manawhenua o Aotearoa me ōna Kōrero Tūturu programme.

Guideline Framework

The guideline outlines eight steps for preparing risk management plans for heritage places. A brief outline of each step and their corresponding chapter of the guideline are discussed as follows.

Steps 1–4 and 8 are envisaged to be undertaken by the owners and managers of heritage places, whereas Steps 5–7 will require specialised inputs.

Step 1: Understanding Risk and Risk Management Planning

This step provides a brief overview of risk and risk management. It identifies risk as a function of hazard, exposure, and vulnerability. It describes risk management as a process of identifying, assessing, managing, and preparing for any risk that could affect heritage places. This chapter describes the disaster management cycle; in New Zealand's disaster risk management framework, the cycle is defined by four terms: reduction, readiness, response, and recovery (4 Rs) (see Figure 2.3). The step introduces the planning process (see Figure 2.2) and the eight steps of the guideline.

Step 2: Developing a Framework for Managing Risk

This step includes advice on establishing organisational commitment, roles, and responsibilities and assessing the operating environment. The guideline considers everyone who may be responsible for managing heritage places from individuals to large organisations. This step aims to promote methods to embed risk management into the daily management of heritage places.

Figure 2.3 Disaster risk cycle.

Source: Heritage New Zealand Pouhere Taonga, 2022

Step 3: Communication and Engagement

This step promotes the early identification of and continued engagement with stakeholders including, Treaty partners, kaitiaki or custodian communities, specialists, local authorities, Civil Defence, funders and sponsors, and community groups. To emphasise the importance of stakeholder engagement in heritage risk management, this was included as an early step in the guideline.

Step 4: Developing Risk Management Policy and Establishing Risk Criteria

This step encourages heritage owners and managers to create a risk management policy in accordance with the owner or organisational goals and objectives, risk management outcomes and timelines, accountabilities, roles, and responsibilities, for resourcing, managing, and reporting risk.

This step identifies legislation that will influence the development of risk management policy. It also advises on establishing risk criteria to evaluate the significance of risk to a place and inform decision-making.

Once the risk management policy and criteria are developed, the owner or organisation will be equipped to identify the need for additional expertise. It is envisaged that this step will inform project briefs for engaging specialists for Steps 5–7.

Step 5: Understanding the Place and Physical Context

This step provides guidance on identifying and understanding the heritage values of a place, it encourages engaging stakeholders to achieve this. It also covers gathering information on the place's physical context including the site's boundaries and buffer zones, mapping, recording the physical condition, and collecting baseline data and sources of information to assist with this process.

Step 6: Understanding the Risk

This step outlines the risk assessment process. The chapter covers identifying risks specific to the place, past disasters, sourcing information, risk analysis, producing hazard summaries, a hazard scape, and a risk evaluation.

This step aims to assist owners and organisations in understanding each hazard, its potential risk to their heritage place, and prioritising the risks to be managed or treated. This step is intended to assist with cost-benefit analysis and the preparation of prevention and mitigation strategies so that actions can be incorporated in to work plans and be budgeted for. In tandem with the preceding steps, this step also promotes stakeholder engagement.

To analyse risk, the chapter proposes the use of hazard summaries which facilitates consideration of the following:

- Hazard overview
- Hazard likelihood
- Exposure and vulnerability
- Hazard consequence
- Impact on heritage values
- How to manage that specific risk?
- What more could be done?
- What is the future risk?

Risks posed by each identified hazard can then be evaluated using a risk evaluation matrix to assign risk ratings and identify the most significant risks to the place (See Figure 2.4).

LIKELIHOOD	CONSEQUENCE: 1 INSIGNIFICANT	2 MINOR	3 MODERATE	4 MAJOR	5 CATASTROPHIC
A: ALMOST CERTAIN OR CUMULATIVE	Medium	High	Very High	Extreme	Extreme
B: LIKELY	Low	Medium	High	Very High	Extreme
C: POSSIBLE	Low	Medium	Medium	High	Very High
D: UNLIKELY	Very Low	Low	Medium	High	Very High
E: EXTREMELY UNLIKELY	Very Low	Very Low	Low	Medium	High

Figure 2.4 Risk evaluation matrix.

Source: Heritage New Zealand Pouhere Taonga, 2022

The terms used to describe likelihood in the matrix denote the following[14]:

A: Almost certain or cumulative – is expected to occur in most circumstances
B: Likely – will probably occur in most circumstances
C: Possible – might occur at some time
D: Unlikely – could occur at some time
E: Extremely unlikely – may occur only in exceptional circumstances

The terms used to describe consequence in the matrix denote the following[15]:

1: **Insignificant** – no injuries, little or no damage, low financial loss, insignificant impact on cultural heritage values
2: **Minor** – first aid treatment required, minor building damage, medium financial loss, minor adverse impact on cultural heritage values
3: **Moderate** – medical treatment required, moderate building and infrastructure damage, high financial loss, moderate adverse impact on cultural heritage values
4: **Major** – extensive injuries, high level of building and infrastructural damage, major financial loss, major adverse impact on cultural heritage values
5: **Catastrophic** – deaths, most buildings extensively damaged and major infrastructural failure, substantial financial loss, and irretrievable loss of cultural heritage values

Step 7: Treating and Managing Risk

This step focuses on the 4 Rs of risk management in New Zealand – reduction, readiness, response, and recovery. It outlines how the risks identified in Step 6 should be considered under each stage of the disaster risk management cycle, the emphasis is on reducing risk and being ready.

The output of this step is a set of action plans setting out priorities, actions, cost, timeframe, and who is responsible over the short, medium, and long term. The approach enables consideration of options that best meet an owner or organisation's risk management objectives, priorities, and available resources. The guideline also provides examples of reduction, readiness, response, and recovery actions. The guideline emphasises the need to balance the costs and benefits of reducing risk and the likelihood and consequence of potential events. It highlights the importance of developing implementable and achievable actions that address risks specific to the place.

Step 8: Implementation, Monitoring, and Review

This step advocates for the need to monitor and review the implementation of the action plans and risk management processes.

It encourages monitoring and updating the risk management plan based on the lessons learnt and actions undertaken. In addition, it recommends a regular review cycle to assess the effectiveness of actions implemented to protect the place.

The end of the guideline includes a glossary of terms adapted by the working group from Disaster Risk Management and heritage models to the New Zealand context, along with a list of further guidance and references.

Conclusion

This chapter has documented the experience of HNZPT in developing a guideline to assist owners and managers of New Zealand's heritage places to prepare risk management plans. The chapter also outlines the eight-step guide to creating a risk management plan developed from that experience.

The experience demonstrated that preparing a risk management plan for a heritage place proved invaluable to the working group for developing the guideline. This offered the group practical experience in adapting international disaster risk management frameworks and models to New Zealand's existing risk and emergency management system. The experience also highlighted the importance of using the language and framework established by New Zealand's National Emergency Management Agency to communicate and align the process for heritage places.

The working group recognised that effective risk management for heritage places requires the engagement of a wide range of stakeholders. In the past, identification and management of heritage has not necessarily involved stakeholder engagement, this is why the guideline emphasises this and the importance of forming and maintaining existing relationships for effective communication.

The working group also encountered the challenges in transforming disaster risk management concepts, processes, and language into a comprehensive format accessible to a wide range of owners, managers, and stakeholders for whom the guideline was prepared.

New Zealand's recent experience of disasters, the evident effects of climate change, and more recently the COVID-19 pandemic have raised awareness of the need to prepare for disasters. The heightened awareness presents an opportunity to advocate for preparing risk management plans for heritage places and enabled the creation of the HNZPT, guideline for preparing heritage risk management plans to facilitate that.

Acknowledgements

Principal author of the HNZPT Guidance for preparing risk management plans Calum Maclean, principal authors of the Te Pitowhenua/Waitangi

Treaty Grounds risk management plan Bill Edwards and Calum Maclean, fellow working group members: Jonathan Howard, Barbara Rouse, Xavier Forde, Sherry Reynolds, Alison Dangerfield, Francesca Bradley, and Dave Margetts (Heritage New Zealand Pouhere Taonga). And special thanks to Catherine Forbes, Helen McCracken, Amanda Ohs, Carole-Lynne Kerrigan (ANZ ICOMOS Disaster Risk Preparedness Working Group), and Richard Nester (Department of Conservation).

Notes

1 Te Ara: The Encyclopedia of New Zealand (2010). *Canterbury (Darfield) Earthquake – Te Ara Encyclopedia of New Zealand.* https://teara.govt.nz/en/historic-earthquakes/page-12. Accessed January 6, 2021.
2 Ibid.
3 Forbes, C. (2019). Are We Really Prepared for Disaster? Responding to the Lessons From Christchurch. *Historic Environment*, 31(2). Conserving It Together – Heritage at Risk, Australia ICOMOS. https://australia.icomos.org/wp-content/uploads/Are-we-really-prepared-for-disaster.-Responding-to-the-lessons-from-Christchurch-vol-31-no-2.pdf.
4 Ibid.
5 Such as the Sendai Framework, 2015, for example.
6 Heritage New Zealand Pouhere Taonga Act 2014 s81(2)(b).
7 Heritage New Zealand Pouhere Taonga Act 2014 s82(4)(c).
8 Waitangi National Trust (2019). *Risk Management Plan Te Pitowhenua / Waitangi Treaty Grounds*. Unpublished Report.
9 Ministry of Civil Defence and Emergency Management (2018). *CDEM Group Planning: Director's Guideline for Civil Defence Emergency Management Groups [DGL 09/18]*. MCDEM. https://www.civildefence.govt.nz/assets/Uploads/logistics-dgl/DGL-09-18-CDEM-Group-Planning-PDF.pdf.
10 Standards New Zealand (2009). *AS/NZS ISO 31000:2009 Risk Management – Principles and Guidelines*. Standards New Zealand. https://www.iso.org/standard/43170.html.
11 Civil Defence Northland Emergency Management Group (n.d.). *Paihia Northland, Community Response Plan*. https://www.nrc.govt.nz/civildefence/community-response-plans/.
12 UNESCO Amman Office (2012). *Risk Management at Heritage Sites: A Case Study of the Petra World Heritage Site*. Paris: UNESCO, p. 14.
13 Ministry of Civil Defence and Emergency Management, *CDEM Group Planning*.
14 Ibid., p. 68.
15 Ibid., p. 69, adapted to include cultural heritage values.

References

Civil Defence Northland Emergency Management Group. (n.d.). *Northland Civil Defence Emergency Management Plan 2016–2021*. Unpublished Report. www.nrc.govt.nz/media/21cfvyys/northlandcivildefenceemergencymanagement-plan20162021.pdf.

Forbes, C. (2019). Are We Really Prepared for Disaster? Responding to the Lessons from Christchurch, in Australia. *ICOMOS Historic Environment*, 31(2).

Heritage New Zealand Pouhere Taonga. (2022). *Guidance for Preparing Heritage Risk Management Plans*. Heritage New Zealand Pouhere Taonga. https://www.heritage.org.nz/resources/-/media/fe8f69a4fd084abea059ca667a7e0537.ashx.

Ministry of Civil Defence and Emergency Management. (2018). *CDEM Group Planning: Director's Guideline for Civil Defence Emergency Management Groups [DGL 09/18]*. MCDEM. www.civildefence.govt.nz/assets/Uploads/logistics-dgl/DGL-09-18-CDEM-Group-Planning-PDF.pdf.

Standards New Zealand. (2009). *AS/NZS ISO 31000:2009 Joint Australia New Zealand International Standard Risk Management – Principles and Guidelines*. Standards New Zealand. https://www.iso.org/standard/43170.html.

The Te Ara the Encyclopedia of New Zealand. (2010). *Canterbury (Darfield) Earthquake – Te Ara Encyclopaedia of New Zealand*. https://teara.govt.nz/en/historic-earthquakes/page-12. Accessed 6 January 2021.

The Te Ara the Encyclopedia of New Zealand. (2011). *Christchurch Earthquake – Te Ara Encyclopaedia of New Zealand*. https://teara.govt.nz/en/historic-earthquakes/page-13. Accessed 6 January 2021.

UNESCO Amman Office. (2012). *Risk Management at Heritage Sites: A Case Study of the Petra World Heritage Site*. Paris: UNESCO.

UNISDR. (2015). *Sendai Framework for Disaster Risk Reduction 2015–2030*. Geneva: UNISDR.

Waitangi National Trust. (2019). *Risk Management Plan Te Pitowhenua/Waitangi Treaty Grounds*. Unpublished Report.

3 Fire Risk Mitigation Strategies at Urban Heritage Sites

Abdelhamid Salah al-Sharief

Introduction

Most Egyptian urban heritage sites are suffering nowadays, especially Historic Cairo, listed by UNESCO as a World Heritage Site (WHS) in 1979. Consequently, the city is exposed to challenges like real estate and building speculation, pollution, and deterioration.

These challenges were exacerbated during the four years following January 25, 2011 revolution. Lack of security and chaos the country experienced allowed real estate speculators to raid the historical areas, irreversibly destroying the traditional urban fabric. In the purview of these facts, it is imperative that Historic Cairo needs a disaster risk management plan to protect cultural heritage and prevent risk effectively as a tool for adequate protection of cultural heritage and prevention of risk. In addition, raising awareness among the community, professionals, and experts to set up a list of priorities to identify the city's real needs is essential.

The chapter focuses on developing an integrated strategy for fire risk mitigation for Historic Cairo and the selected study areas of al-Azhar and al-Ghūrī. The proposed mitigation measures are tailored to match the needs of the study area to be used as a prototype for all the urban districts of Historic Cairo.

The structure has been developed based on four pillars to fulfil the main objective of proposing a tailored fire risk mitigation. The first pillar is the value assessment of the selected case study. The second pillar is evaluating the conservation state of the built heritage. The third pillar is proactive assessment for providing sufficient information that identifies and allocates potential ignition sources. The fourth pillar is fire risk mitigation which aims to provide integrated measures tailored to the requirements of the study area to minimise the fire risk and its level of impact on the cultural heritage value.

Significance of the al-Azhar and al-Ghūrī District

Many discussions took place to identify the historical sites and determine their properties. According to Stovel (2007), authenticity and integrity are the main

DOI: 10.4324/9781003356479-4

factors in assessing a historical city where wholeness refers in historic cities to all the characteristics shared in the outstanding universal values (OUV) of the city.

Silva and Roders (2012) add that managing cultural heritage primarily concentrates on cultural significance transferred through values and characteristics, including both tangible and intangible.

In other words, the significance of a historical city could be determined by properties related to the city's location, design, history, quantity and condition of its buildings, the inhabitants and their habits, jobs, and much more.

Al-Azhar and al-Ghūrī district, a rich part of Historic Cairo which represents diverse monuments and a cluster of surviving historical buildings scattered along the site, both as listed monuments, ruins, and derelict, in general, the site is in good physical condition with easy accessibilities. This matches the first criterion of UNESCO's ten criteria for selecting the World Heritage Site nomination.

The area of al-Azhar and al-Ghūrī is characterised by its heritage buildings, especially the two famous landmarks – the al-Azhar and al-Ghūrī complexes – proving architecture to be the most important characteristic of the site. This is based on criterion number (V) of UNESCO.

The site's urban fabric has the highest priority to conserve and protect, the Qasaba street, now called al-Muʿizz street, continued its function from the Fatimid era until now. In addition to the historical buildings listed and not listed as monuments, the streets and the markets all function harmoniously.

Conceptual Framework and Subject Proposition

The selected zone of the al-Azhar and al-Ghūrī inspectorate and its surrounding areas have suffered many risks. According to the inspectors, conservators, and security guards working within the selected inspectorate, the earthquake that struck Cairo in the late 18th century caused significant destruction. Additionally, numerous fire accidents across the city caused irreversible damage to many historical buildings.

After 2011, the new high-rise buildings replacing the historical houses posed another high risk on the site (see Figure 3.1). This new construction has left the site vulnerable to fire risk, especially the commercial businesses exercised in these buildings, including the illegal buildings, with no preparedness and measures to fight fires.

Hence, the Historic Cairo zone is selected considering its dire need for developing and integrating prevention and mitigation strategies to reduce the fire risks and its impacts on the urban heritage and rising vulnerabilities to different hazards such as earthquakes.

Figure 3.1 New buildings around historical buildings.

Source: Author

Proposed Fire Risk Assessment Methodology (FRA)

Recently, threats to heritage properties have increased due to fire risks. The severe impact of fire on built heritage and the urban fabric prioritises Fire Risk Assessment (FRA) followed by Fire Risk Mitigation (FRM) in any conservation plans.

The al-Azhar and al-Ghūrī districts are vulnerable to fire due to the sensitivity of heritage materials, traditional urban fabric, and improper human and commercial businesses lacking fire safety measures.

The proposed fire risk assessment methodology depends on identifying the factors affecting and increasing the impact of fire risk at the site. The proposed methodology shall be applied using two parallel methods (see Figure 3.2). Fire risk has the highest destructive impact on the Historic Cairo monuments and buildings, resulting in total loss or partial damage. The study area begins from al-Azhar street in the north to al-Muʿizz street in the east, al-Azhar Park in the west, and Bab Zuweila in the south.

Condition Assessment of al-Azhar and al-Ghūrī

Information from the site was collected to assess the study area for recording the deterioration aspects of the site.

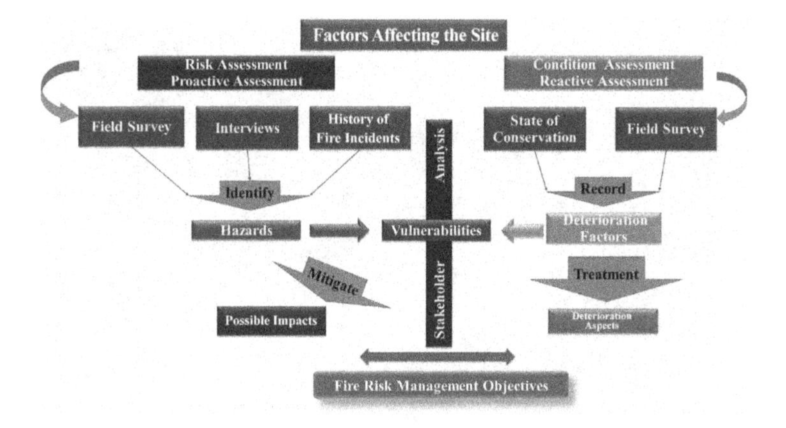

Figure 3.2 Proposed fire risk assessment methodology for fire mitigation for the selected site.

Source: Author

State of Conservation of the Site

Evaluating the conservation state of the site is an important factor in the methodology connected to fire ignitions and propagation. Generally, the evaluation of construction state, including listed monuments, historical buildings, and commercial and residential activities, focuses on evaluating both electrical and gas installations. The study allocated the activities dependent on gas containers on a map and their distance to the historical buildings. None of those activities has any fire safety measures installed.

Overall Condition of the Built Heritage

The overall condition of the structure varies from extreme damage to medium. Less than 10% of the monuments and historic buildings are in poor condition due to prominent factors like soil subsidence, subsurface water, and deteriorating wooden ceilings, windows, and doors are also dilapidated, allowing waste and dust accumulation (see Figure 3.3). The heaps of waste, especially solid waste, are considered extreme sources of fire propagation.

The study zone has 1,100 open stores and warehouses, of which around 40% are fabrics and cloths markets, and around 11% are the book market. The groceries, meat, and vegetable stores are around 11%, and cafés, restaurants, and bakeries are around 9%. Furniture and traditional crafts workshops, such as inlaid wooden artefacts and leather products, are around 8%.

Figure 3.3 Demolition of old buildings, Atfit Hosh el-Nimr.

Source: Author

Risk Assessment

Natarajan and Jayasudha (2019) discuss that urban growth, rapid modification of built shape and environment, and increasing population in the traditional city result in a lack of fire safety measures. Such a situation needs to be systematically analysed to support fire mitigation. The study points out that continuous construction in the old city with inadequate evacuation measures in emergency times increases the risk of disaster during a fire. The study also states that fire ignition in the urban infrastructure is connected to parameters, like space between buildings and the state of the constructions (Natarajan and Jayasudha, 2019).

The United Nations for Disaster Risk Reduction (UNSDRR), within its handbook about local government leaders (2017), states new methods for fire risk mitigation and emergency response to fire based on the Manchester August 2013 fire accident. The first method states involving consultants to develop conditions to provide safety procedures to be applied in planning for grant permission. Secondly, both site owners and public authorities shall coordinate to reduce fire risk by promoting safe working practices. Thirdly, learning from the ex-fire accidents at the waste transfer site (Gencer, 2017; UNISDR, Geneva, 2017 Version).

Twigg (2017) stated that buildings set closely together in urban settings are more exposed to the threat of fire, especially the informal and unplanned settlements that result in a lack of infrastructure and poor services. This is mainly attributed to the lack of holistic and precise data about the impact of fire and fire incidence. Hence, improving data is vital to conducting a risk assessment to develop relevant mitigation and prevention strategies and preparedness plans for responding to fire. Besides, decision-takers must be aware of the history of fire incidents and the level of fire risk in urban settlements (Twigg et al., 2017).

Therefore, this element of the proposed methodology demonstrates the three aspects of the history of fire accidents: identifying ignition sources, recording the study area's vulnerabilities, and accelerating the effect of fire risk.

History of Fire Incidents Within the Study Area

The study area was selected based on the history of fire incidents in the area, its close quarters, and the high probability of the potential fire risk to the site.

As per records, in the last ten years, fire has caused a destructive effect on monuments and historical buildings in Historic Cairo in terms of total loss and partial damage.

A few noted incidents are as follows:

- In June 2016, a fire engulfed *Atfit Hosh Nimer* causing two historical buildings to collapse.
- In November 2016, a fire erupted in vacant land in *Atfit el-Sokary.*
- In May 2017, a fire destroyed two historical buildings in *Haret al-Sayad.*
- In August 2016, a fire broke out at a warehouse for painting materials.
- In October 2016, a residential apartment in *Haret Hosh Kadam* caught fire.
- In February 2017, one of the shops in *Haret el-Madrasa* also caught fire.

Besides these, many other fire incidents caused severe destruction to the urban fabric of the site and its surrounding areas.

The Source of Ignition or Fire Hazard

It is inferred from the overview of the history of fire incidents and the condition assessment of the study area that the al-Azhar and al-Ghūrī zone are prone to fire incidents. The level of damage due to fire, potential ignition source, and fire propagation speed are connected with vulnerabilities in the area despite its location inside the buildings.

Fire ignitions are linked to vulnerabilities such as the state of conservation, electricity, gas installations, and improper residential and commercial businesses. On the other hand, the speed and level of fire propagation rely on many factors, like the degree of flammable elements within the buildings, their combustibility, speed, and flammability. Additionally, the vacant land within the urban fabric is another important factor, including solid waste with combustible properties, fire preparedness measures, and inadequate water and sewage systems.

A fire hazard can be a primary hazard based on the source of ignition and vulnerabilities. The diagram in Figure 3.4 represents correlations between the ignition sources and vulnerabilities within the study area, which also help to decide the best level for intervention to prevent fire propagation.

Moreover, fire can be a secondary hazard induced by another primary hazard. Primary hazards like earthquakes can lead to fire as a secondary hazard as per the vulnerabilities recorded during the field survey. Fire can also result from heavy rain, causing a risk of water damage due to insufficient sewage or water systems. In the last nine years, the study area witnessed three terrorist attacks. These terrorist attacks are also a primary hazard causing fire, a secondary hazard due to the vulnerabilities recorded at the site.

The lack of security at the registered monuments encourages residents to dispose of waste beside the historical buildings and inside them. Furthermore, the unoccupied demolished houses are another area to throw waste. Due to this practice, the accumulated waste becomes another source of fire risk.

Nearly 90% of the residents and shop owners dispose of waste on the building's roof. Over the years, this waste dump has reached a level that might lead to a major fire since it is an unattended space.

According to the field survey, the study area has inadequate fire response and mitigation strategies due to the following reasons:

- Overpopulation and congested urban fabric with narrow to no gaps between the buildings accelerate fire propagation.

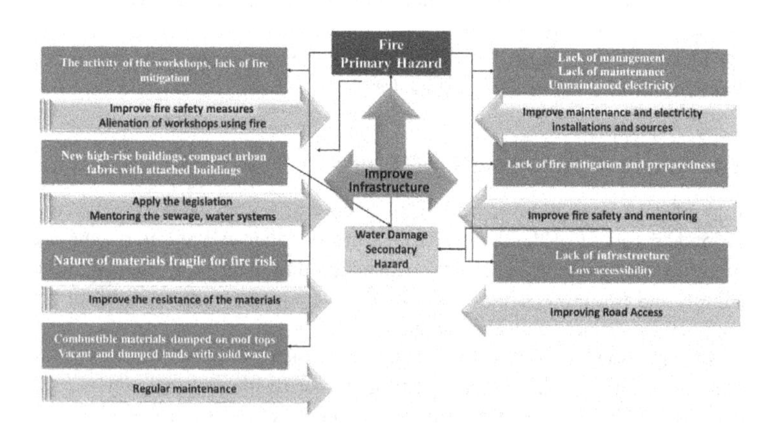

Figure 3.4 Mitigation measures for fire primary hazard based on the risk assessment.

Source: Author

- Lack of accessibility of the zone; the streets are too narrow for fire trucks to pass through, complicating the firefighting process and increasing response time.
- Majority of vacant lands are informally used for waste collection. This causes severe vulnerability speeding up the need for fire propagation due to flammable materials found in waste. Besides, this waste is fuel for the fire, especially given certain circumstances when it becomes self-combustible.
- Increasing commercial businesses alongside residential places increases the fire risk, especially with some businesses using fire for production and services.
- Rapid urbanism with illegal high-rise buildings in narrow streets adds complexity to firefighting, especially if the fire strikes on the upper floors.
- Most commercial businesses lack efficient fire protection measures and mitigation measures.
- Commercial activities usually use a part of the street in front of shops to display some goods and equipment. As a result, the fire response time can be delayed as a cause of these barriers.

Given these facts, historical buildings most at risk of fire have been identified and marked on the city's map.

Proposed Fire Risk Mitigation (FRM) Measures

Mitigation strategies are fundamental when prevention measures are complex and unable to achieve their objectives. FRM strategy should encompass all possibilities of fire risk, including ignition source, fire propagation, consequences of a fire incident, and impacts due to fire. Additionally, it must consider the types of fire hazards, primary or secondary, for continuous risk assessment.

Fire – as a Primary Hazard

The risk assessment concluded that fire, as a primary hazard, results from different ignition sources mainly due to inadequate fire safety measures in commercial and residential activities. Furthermore, it is linked to various vulnerabilities existing within the heritage properties and the urban fabric.

The main mitigation measures for fire as a primary hazard depends on reducing the vulnerabilities of the concerned area inside heritage proprieties and within the urban fabric.

Fire – as a Secondary Hazard

According to the risk assessment, fire, as a secondary hazard, can result from three primary hazards such as seismic shock, water damage, and terrorist attacks. The risk analysis concluded that fire hazard is generated from these primary hazards due to the recorded vulnerabilities of the site.

The mitigation measures are similar and rely on reducing the vulnerabilities, especially those that can generate fire and speed up its propagation.

On a large scale, some significant measures need to be implemented to mitigate fire risk for both the short and medium term.

The Following Steps Represent the Proposed Short-Term Strategy

- Eliminating unnecessary ignition sources by
    - Removing all garbage and solid wastes from the zone.
    - Separating garbage sources from combustibles.
    - Managing and using safe equipment inside heritage sites and workshops and raising awareness for using simple fire preparedness such as fire extinguisher.
- Limiting contact with combustible materials, especially around heritage sites.
- Vegetation management to limit fire propagation.
- Implementing fire mitigation and preparedness procedures involving stakeholders.
- Providing heritage sites with manual suppression equipment.

The Following Measures Represent the Proposed Mid-Term Strategy

- Developing a plan to use the vacant land for heritage facilities.
- Upgrading electrical wiring in heritage sites and workshops along with implementing regular maintenance.
- Removing stoves from historical buildings and providing a secured remote building for workshops.
- Training security to assist in limiting potential intentional fires.
- Rearranging electrical needs to accommodate the use of the commercial shops to reduce the ignition sources.
- Moving electrical transformers away from any flammable materials.
- Proposing a fire detection system that can be maintained according to the capacity of the current management system of the stakeholders.
- Applying suppression systems or suitable fire preparedness equipment that are applicable to the different sources of fire, for example, the water cannot be used to fight the fire generated by electricity problems..

Applying Method for Mitigating Fire Risk in the Study Area

Considering the risk assessment and the proposed strategy for fire risk mitigation, the primary measure of the mitigation by blocking the risk path to prevent fire hazards from reaching the sites was applied (see Figure 3.4).

The Processes of the Applied Method

The criteria for prioritising the interventions were set according to the degree of risk to the selected monuments and organising the work periodically between sites to ensure the desired work goal continuity. An agenda was formulated for coordinating with concerned ministries to target nearby archaeological sites from each other to undertake periodic cleaning and waste disposal for fire risk prevention. This applied method would result in optimal benefit for many monuments.

The intervention took place between April 2018 and December 2020 to achieve optimal benefit from the applied method. The rate of solid waste accumulation, an ignition source of fire, within and around the targeted monuments was studied to identify the accumulative level of each for preparing a working schedule with set periods for waste removal from inside the building and the surroundings of each site. Additionally, the method helped conserve the targeted monuments in an appropriate and safe condition for visitors and inhabitants. According to the availability of workers, a schedule is set up for periodic monthly maintenance and quality check for work progress across all targeted sites.

Outcomes of the Applied Method

As a result of the previously applied method, there were many important benefits for the site layout, people, and monuments which could be concluded as follows:

- Monitoring and marking all streets in the indicated area defined on the map.
- Identifying the target monuments and marking them on the map.
- Identifying the vacant lands used as informal waste collection places on the map and measuring the volume of waste in cubic metres.
- Determining all fixed and temporary obstacles in the streets of the study area that result from the commercial activities and marking them on the map.
- Locating fire hydrants and fire hoses on the map.

- Locating electricity boxes and telephone lines on the map.
- Determining commercial businesses constituting a potential source of fire and its proximity and propagation speed to the historical building to determine the source of waste.
- The accessibility to the sites and historical buildings became simple due to measuring the street breadth and connecting them to the main accesses to the ancient buildings from outside. The availability of such information undoubtedly enhances the collaboration between the Civil Defence Department and the Ministry of Tourism and Antiquities to the rapid response of civil defence vehicles from outside the district to inside them. Moreover, both became aware of the capability proportionality available to each to develop plans to prepare for the fire risk.
- Marking the monuments exposed to fire risk, either due to population or commercial businesses that represent a source of the fire.
- Identifying waste collection sites which would increase weaknesses and increase the fire propagation speed from its source to the targeted monuments.
- Monitoring population activities and commercial businesses that increase fire risk due to their activities and marking them on the maps.

Conclusion

Historic Cairo is the only urban WHS in Egypt, with various monuments, cultures, and communities for more than 1400 years. An Islamic capital with its cultural heritage along with the Egyptian generations holds the true assets of Egypt. The cultural heritage from monuments, sites, and artefacts with a historic timeline provides the best example of diverse human culture.

Fire hazard, or anthropogenic hazards, in general, proves that fire incidents adversely damage the cultural heritage. The need to formulate and develop disaster risk management plans with integrated mitigation strategies and tailored preparedness plans has become an urgent need for heritage conservation.

It is crucial to share experiences and collaborate with national and international institutions to unify efforts and ensure effective plans are prepared to manage cultural heritage during a crisis.

Producing manuals, guidelines, and training with updated modules to the local authorities and organisations is imperative to qualify intervention teams to manage fire incidents and minimise cultural heritage damage.

The required procedures and aspects of developing an integrated fire risk mitigation plan for the study area like Historic Cairo must consider the urban and living heritage challenges involving many stakeholders.

The project formulated criteria corresponding to the requirements of various cultural heritage and specific urban and regional policies throughout many aspects. Fire risk is threatening the unique urban fabric of Historic Cairo as both a primary and secondary hazard. Fire has caused severe damage to historical buildings, so the project provides a tailored methodology for fire risk assessment for the urban heritage applicable to each quarter of Historic Cairo. The project proves that fire risk mitigation can be implemented with a low budget when fire risk assessment is conducted scientifically and methodologically. Furthermore, it helps to identify the stakeholders and assess their financial capabilities and technical recourses.

The project formulates an integrated mitigation strategy on a national level. The strategy is divided from short to long term by planning preventive measures, including preparedness, adequate response, and recovery measures.

The project set up a tool that can be replicated in other heritage sites in Egypt. At the same time, it will help the trained personnel to enhance their skills and capacity.

Lastly, the study presents a set of actions and processes to be performed by the involved on-ground stakeholders before, during, and after a fire incident.

References

Behrens-Abouseif, D. (1989). *Domestic Architecture in Cairo. In Islamic Architecture in Cairo: An Introduction*. New York: E.J. Brill.

Behrens-Abouseif, D. (2007). *Cairo of the Mamluks: A History of the Architecture and Its Culture*. Cairo: The American University in Cairo Press.

Creswell, K.A.C. (1919). *A Brief Chronology of the Muhammadan Monuments of Egypt to A.D. 1517*. Cairo: IFAO.

Creswell, K.A.S. (1978). *The Muslim Architecture of Egypt*, vol. I–II. New York: Hacker Art Books.

El-Basha, H. (1970). *Cairo Its History, Arts and Monuments*. London: J. S. Virtue & Company.

El Zafarany, A.M. (2011). *Urban Regeneration Project for Historic Cairo Sector Study: Environmental Risks Facing Historical Cairo*. Report. URHC. https://www.urhcproject.org/Content/studies/2_zafarany_environmental.pdf.

Ferreira, T.M., Vicente, R., da Silva, J.A.R.M., Varum, H., Costa, A., and Maio, R. (2016). Urban Fire Risk: Evaluation and Emergency Planning. *Journal of Cultural Heritage*, 20.

Gencer, E. (2017). *How to Make Cities More Resilient: A Handbook for Local Government Leaders*. Geneva: UNISDR.

Jigyasu, R. (2005). *Towards Developing Methodology for Integrated Risk Management of Cultural Heritage Sites and Their Settings*. Conference Session. 15th ICOMOS General Assembly and International Symposium, Xi'an, China.

Jigyasu, R. (2011). *Lecture on Disaster Risk Mitigation for Cultural Heritage: FAC Course*. ICCROM. https://www.iccrom.org/courses/disaster-risk-management-cultural-heritage-3.

Natarajan, R.B., and Jayasudha, P. (2019). Fire Risk Mitigation at Urban Scale in Commercial Buildings of Kumbakonam Town. *International Journal of Recent Technology and Engineering*, 8(3), 2920–2928.

Silva, A., and Roders, A. (2012). *Cultural Heritage Management and Heritage (Impact) Assessments*. Conference Session. Joint CIB W070, W092 & TG International Conference: Delivering Value to the Community. https://scholar.google.com/citations?view_op=view_citation&hl=en&user=c5T-fE8AAAAJ&citation_for_view=c5T-fE8AAAAJ:qjMakFHDy7sC.

Stovel, H. (2007). Effective Use of Authenticity and Integrity as World Heritage Qualifying Conditions. *City & Time*, 2(3), 3.

Twigg, J. (2015). *Disaster Risk Reduction*. London: Overseas Development Institute.

Twigg, J., Christie, N., Haworth, J., Osuteye, E., and Skarlatidou, A. (2017). Improved Methods for Fire Risk Assessment in Low-Income and Informal Settlements. *International Journal of Environmental Research and Public Health*, 14(2), 139.

UNESCO. (2012). *Urban Regeneration Project for Historic Cairo*. Paris: UNESCO.

UNESCO, World Heritage Convention (2005). *Operational Guidelines for the Implementation of the World Heritage Convention*. Paris: UNESCO.

(ت.930هـ/1523م)، بدائع الزهور في وقائع الدهور، تحقيق محمد مصطفى، الجزء4، القاهرة، الهيئة (محمد بن أحمد بن اياس الحموى) ابن اياس المصرية العامة للكتاب، 1984.

المقريزى المواعظ و الاعتبار بذكر الخطط والأثار، مطبعة الادب، القاهرة 1968.

حسن عبد الوهاب، تاريخ المساجد الأثرية، بيروت، 1993، ص ص286–294.

رحلة مع أسبلة القاهرة، القاهرة التاريخية. المجلس الأعلى للأثار، القاهرة.

1980 3، ،وزارة الأوقاف، المجلس الأعلى للشئون الإسلامية، القاهرة سعاد ماهر، مساجد مصر وأولياؤها الصالحون، ج

4 Ancient Town Walls at Risk

Methods, Technologies, and Tools for Multi-hazard Risk Analysis, Monitoring, and Governance

Francesca Giuliani

Introduction

The chapter presents research activities undertaken for ancient town walls, which were partly developed within the project Monitoraggio delle Mura Urbane (MO.M.U.) for monitoring town walls. The project aims to innovatively contribute towards a planned, programmed conservation of the ancient town walls, focusing on defining methods, technologies, and tools for their multi-hazard risk analysis, monitoring, and governance.

Recent damage to the historical town walls in the Tuscany region, specifically in the settlements of Volterra (2014), Magliano in Toscana (2012), Pistoia (2011), Cana di Roccalbegna (2013, 2020), Poggio a Caiano (2017), and San Gimignano (2018), has raised attention to the growing vulnerability of the heritage-listed fortifications. These ancient assets were historically significant for developing communities and defending people and places from foreign invaders since the remotest period. The extension and consolidation of conquests and the continuous conflicts with nearby cities led to the design and construction of advanced defensive systems that are considered monuments of inventive genius and skill (Toy, 2006). Although they have progressively lost their protective function, fortifications, particularly town walls, constitute large preserved iconic attributes of civic pride and identity.

Beginning in 2019, the Regional Government of Tuscany promotes systemic research regarding the walled historic centres to develop a fit-for-purpose methodological framework to inform risk governance in its territory. The effective implementation of the planned preventive conservation of these wide monumental constructions, spread across a vast territory, is related to Heritage Asset Management (HAM). Even today, a standard process or framework for HAM is non-existent. However, there is a general agreement on the necessity to ensure systematic, condition and significance-based conservation, repair, and maintenance. Creating a decision support system for the Tuscan town walls is based on the multidisciplinary and cross-scalar knowledge of all

DOI: 10.4324/9781003356479-5

regional assets that are included in a Geographical Information System (GIS) based platform. The conception and implementation of novel tools for managing the conservation process may increase the operative capacity by fostering the transfer and sharing of data between different governmental stakeholders, from regional to local. The proposed framework promotes decision-making based on knowledge and comprehensive and updated data.

Until now, more than 140 ancient walled systems have been singled out and mapped into a regional GIS platform (see Figure 4.1), collecting and georeferencing several general information, such as the history, construction features, archival material, historical images, and recent photographs (De Falco et al., 2022). Even though they present varied configurations and morphological features, they are primarily made of brick or stone multi-leaf masonry, whose composition, geometry, and arrangement depend on local materials and workmanship. Most of the walled systems date back to Medieval times, in some cases incorporating the ancient defensive systems, particularly the Etruscan or Roman. Only a few cases present modern fortifications built in the second half of the 15th century. Medieval walled systems are composed of isolated slender walls connecting high-rise towers, whereas modern walls present thicker scarp-shaped sections and bastions, with terreplein behind the walls. Today, the town walls are deeply integrated into the urban fabric, and their accessibility can be difficult since many constructions, even private housing units, have been progressively reusing their structural elements.

This chapter presents the preliminary results of several activities regarding the town walls in Tuscany, ranging from territorial to local-level investigations. The application of invasive tests was not feasible due to the heritage value and the impossibility of altering the physical aspect. Hence, digital

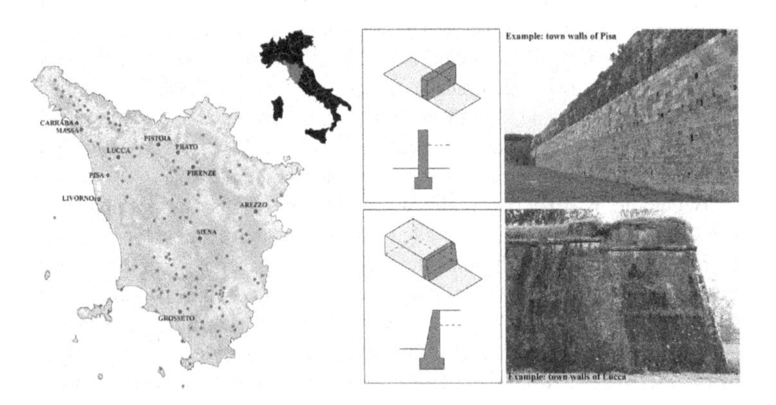

Figure 4.1 Cities still preserving ancient town walls in Tuscany (Italy).

Source: Author

non-invasive technologies were included to support data acquisition, collection, management, and elaboration.

Conceptual and Methodological Framework

Planned preventive conservation entails the protection of cultural heritage and is founded on the attentive identification of risk situations and the systematic planning of minimally invasive interventions (Della Torre, 2021). This approach ensures greater efficiency in terms of costs and results concerning unplanned systems based only on corrective actions (Vandesande and Van Balen, 2018), lacking removal of the causal factors of disasters. Conversely, planned preventive conservation is a proactive management process to avoid unnecessary deterioration, damage, and even failure through periodical monitoring, scheduled maintenance, and integral condition assessment.

In architectural heritage, simplified multi-hazard approaches represent a significant step forward in prioritising risk reduction measures (Giuliani et al., 2021; Ortiz and Ortiz, 2016). Moreover, qualitative index-based methodologies can identify situations when detailed assessments are needed (Romão et al., 2016). The knowledge of materials, technologies, stratifications, and contexts is paramount and can favour a systemic approach. These aspects affect the correctness of risk analysis that requires a critical judgement based on reading and interpretation of the object to be preserved and the phenomena to which it is subjected. Adopting a multi-scale approach to gathering and organising information is recommended since it allows for varying the refinement of investigation and knowledge. Thus, the study of town walls is based on a multi-scale organisation of processes, ranging from the territorial scale to the minute scale of individual elements and masonry materials.

Ancient town walls are affected by various natural and human-induced hazards, whose effects are amplified by climate changes. Thus, a preliminary analysis of hazards and threats affecting the Tuscan sites plays a key role in framing the risk characterisation and methodological framework. Hazards on cultural heritage may be categorised according to the speed of onset (Ravankhah et al., 2019). Sudden-onset events typically cause significant impacts at a structural level (macro-scale), although sequels may be left at a micro-scale. Instead, slow-onset events primarily cause stresses at the material level (meso-micro-scale), even if these are bound to eventually lead to structural collapse if nothing is done to prevent it. This distinction helps practitioners to adequately address the short- and long-term effects of hazardous events on heritage sites. Based on this, a fit-for-purpose index-based methodology (De Falco et al., 2021) has been proposed to assess risks originating from slow-onset hazards, which are the worst-case condition for the town walls. Hence, vulnerabilities are assessed starting from the surveyed degradation and damage. Exposure involves the walled system itself, and the people, objects, buildings, roads, or any element that can potentially be damaged by the failure of the town walls. Five vulnerability, exposure, and risk classes were considered, ranging from low to high values, the latter corresponding to unacceptable or critically acceptable levels.

At the material level, the knowledge of historical masonry structures builds upon the characterisation of two aspects, construction techniques and state of conservation. The first consists of the detection of areas with homogeneous material and texture, usually corresponding to construction phases (Doglioni, 2010). The second groups areas with homogeneous degradation phenomena, such as the presence of vegetation, cracks, rising damp, surface crusts, and spalling of the material. These aspects are commonly annotated on 2D or 3D graphical representations of the structure, among which photogrammetric surveys can realistically reproduce the wall texture. The survey of the historical wall surfaces provides critical information on the capacity of the construction to withstand external actions or, conversely, to estimate the liability to local failure mechanisms, such as disaggregation or crumbling (Borri et al., 2020). The conventional annotation approach is based on the manual drawing of the regions over the images making it lengthy and time-consuming, especially on large monuments like the town walls. Recent research efforts explored the use of AI-powered tools for the semantic segmentation of 2D orthographic images (Pavoni et al., 2020, 2022).

Hazards and Threats

Initially attributed only to extreme climatic events like heavy rainfalls, a closer study of the collapsed walls reveals a complex situation where the weathering conditions of cultural heritage assets contribute to the progressive and slow degradation of building materials (De Falco et al., 2021). For instance, a 25-m-long portion of medieval town wall in Pistoia (see Figure 4.2a) partially collapsed on September 9, 2011. The collapse was triggered by heavy rain on an already critical condition determined by the construction features, like wall slenderness, height, the difference between the ground levels on both sides of the wall, and the poor quality of masonry. In September 2020, the same curtain wall completely collapsed for a length of 30 m without any extreme meteorological event preceding the disaster.

Another emblematic example is the collapse of a tower, 20 m high and 9 m wide, of the medieval town walls of Magliano in Toscana, on November 13, 2012 (see Figure 4.2b). Although an extreme meteorological event triggered the failure, the analysis of the root causes indicated towards human alterations as contributing factors. Shortly before the disaster, the vertical passing-through cracks on the walls had been closed, and all the joints repointed during a consolidation campaign. The closure of these natural drainage holes probably caused the collapse during an exceptional rainfall, when the weep holes at the base were obstructed by soil accumulated in the inner

a) Pistoia (September 19, 2011)

b) Magliano in Toscana (November 13, 2012)

c) Volterra (January 31, 2014)

d) Volterra (January 31, 2014)

Figure 4.2 Emblematic examples of failures of town walls.

Source: De Falco, A., Giuliani, F., Ladiana, D., Rjolli, L., Bordo, D., Gaglio, F., Di Sivo, M.: Typological characterisation of ancient town walls for disaster prevention and mitigation: the MO.M.U. project, in: Roca, P., Pelà, L., Molins, C. (Eds.), 12th International Conference on Structural Analysis of Historical Constructions SAHC 2020, 2021

part of the tower. Another sudden failure occurred on January 31, 2014, in Volterra (see Figure 4.2c and d), where a 35-m-long wall collapsed, after a series of heavy rainfalls that increased the hydrostatic pressure in the sandy terreplein behind the poor-quality masonry wall.

While climate change and natural hazards are seriously affecting historic town walls, even human-related activities and local technological ruptures can potentially cause the collapse of greater sections. Two different hazard conditions can be identified. The first is associated with ordinary conditions and involves several slow-onset hazards, such as pollution, wind erosion, capillarity dampness, and ground subsidence. The second is associated with isolated adverse events, among which landslides, floods, and earthquakes are more frequent in Tuscany.

The Town Walls of Pisa

Benchmarking studies concerned the application of the index-based risk prioritisation procedure and the investigation of materials, either building techniques or degradation phenomena, on the town walls of Pisa. The town walls were built in Medieval times, starting from the 12th century, using local materials, techniques, and workmanship. The overall length is approximately 7 km, the average height of the curtain walls is 11 m, and the mean thickness is 2.20 m. The structure is made of multi-leaf masonry, with two brick and stone outer leaves and the inner core of rubble masonry.

Despite the heterogeneous appearance of the walls, seven prevalent material classes have been identified to characterise locally homogeneous areas showing similar construction techniques (Pavoni et al., 2022).

The material classes consider the lithology, shape and dimension of blocks, the presence of mortar, and their arrangement. The latter accounts for the organisation in coursed rows or radial shapes, the even or uneven height of courses, the presence of snecks, and the way units are overlapped. Among these classes,

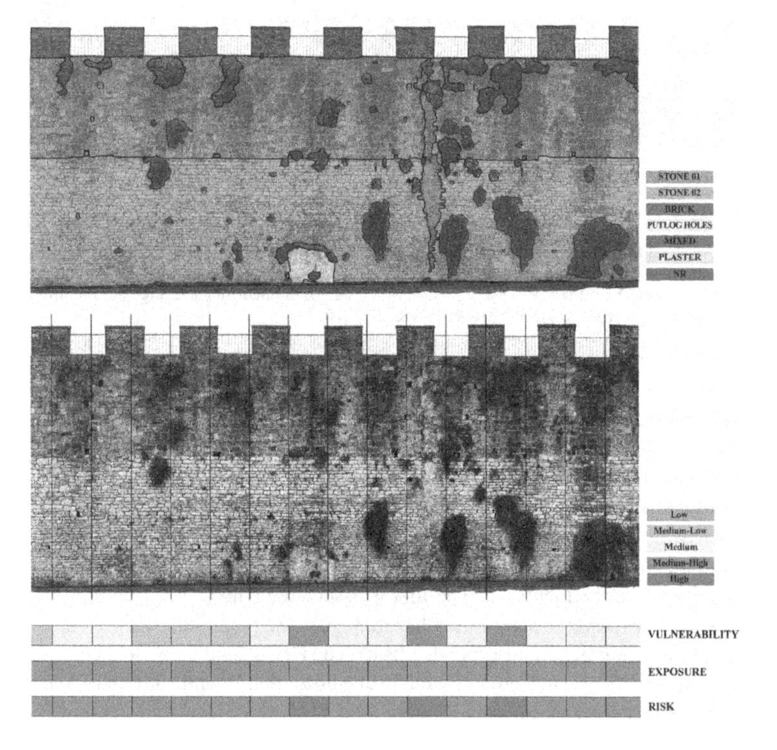

Figure 4.3 Preliminary results of the analysis on the town walls of Pisa. At the top: automatic annotation of masonry performed with TagLab; at the bottom: partition of the town walls and evaluation of risks associated with slow-onset hazards.

Source: Author

one concerns brick masonry, five describe stone walls, and one refers to mixed masonry typical of infilled openings and reconstruction works with diverse materials. Two additional classes have been included to map putlog holes and higher plants, grass, bushes, and even trees that hinder the recording of masonry.

The partial results of the assessment show the orthoimage of a sample portion of the town walls and the results of the annotation process performed with TagLab (see Figure 4.3). The element is divided into 2-m-long units that were evaluated based on vulnerability, exposure, and risks. The coloured stripes at the bottom identify different levels; green squares are characterised by lower scores, while red presents higher scores. Exposure is high due to tourists visiting the town walls and walking on top of the curtain. Conversely, vulnerability is high if cracks, alterations, or strong block erosion are clearly visible. Risks tend to be higher in the proximity of highly vulnerable portions. Hence, monitoring and interventions are recommended to be a priority.

Conclusions and Future Developments

The chapter presented the methodological framework and the research activities conducted on the town walls of Tuscany, Italy. Several procedures and tools were illustrated to provide a comprehensive understanding of risks in this peculiar typology of cultural heritage assets.

By applying this framework to the whole circuit of any walled system, it is possible to obtain digital maps that compare the state of conservation and the risks of diverse portions of the monument, even those located in different areas. These maps are important tools for planning knowledge-based actions and prioritising interventions ensuring appropriate management of economic resources, which are often limited. Specifically, GIS is a powerful tool to create maps, scenarios, and inventories, as users can elaborate a large number of data sets from multiple sources, join hazard, exposure, and vulnerability information, merge inputs from different professionals, and store updated information at any stage of the conservation plan and throughout the monument life cycle.

The research is currently exploring how the application of emerging technologies in remote sensing, such as satellite radar interferometry, can be applied for monument monitoring. Integrating these aspects into a single GIS platform will support the HAM and inform decision-making for risk governance and planned programmed conservation. Lastly, the procedure has the potential for scaling up and adaption to other countries that still preserve this type of cultural heritage, accounting for site-specific needs and conservation strategies.

Acknowledgement

I wish to acknowledge all the people involved in the MO.M.U. project at the University of Pisa, coordinated by Prof. Michele Di Sivo, the Regional

Council of Tuscany, and the LaMMA Consortium. The software TagLab has been developed by the Visual Computing Lab, ISTI-CNR in Pisa, Italy.

References

Borri, A., Corradi, M., and De Maria, A. (2020). The Failure of Masonry Walls by Disaggregation and the Masonry Quality Index. *Heritage*, 3(4), 1162–1198.

De Falco, A., Giuliani, F., Ladiana, D., Rjolli, L., Bordo, D., Gaglio, F., and Di Sivo, M. (2021). Typological Characterization of Ancient Town Walls for Disaster Prevention and Mitigation: The MO.M.U. Project. In: Roca, P., Pelà, L., and Molins, C. (Eds.), *12th International Conference on Structural Analysis of Historical Constructions SAHC 2020*. https://www.iiconservation.org/content/12th-international-conference-structural-analysis-historical-constructions-sahc-2020.

De Falco, A., Gaglio, F., Giuliani, F., & Martino, M. (2022). A BIM-Based Model for Heritage Conservation and Structural Diagnostics: The City Walls of Pisa. In *The Future of Heritage Science and Technologies: Design, Simulation and Monitoring* (pp. 84–96). Cham: Springer International Publishing.

Della Torre, S. (2021). Italian Perspective on the Planned Preventive Conservation of Architectural Heritage. *Frontiers of Architectural Research*, 10(1), 108–116.

Doglioni, F. (2010). Leggibilità della costruzione, percorsi di ricerca stratigrafica e restauro. In: Brogiolo, G.P. (Ed.), *Archeologia dell'architettura*, vol. XV. All'Insegna del Giglio, pp. 65–79. https://intrecci.sira-restauroarchitettonico.it/index.php/intrecci/article/view/16.

Giuliani, F., De Falco, A., Cutini, V., and Di Sivo, M. (2021). A Simplified Methodology for Risk Analysis of Historic Centers: The World Heritage Site of San Gimignano, Italy. *International Journal of Disaster Resilience in the Built Environment*, 12(3), 336–354.

Ortiz, R., and Ortiz, P. (2016). Vulnerability Index: A New Approach for Preventive Conservation of Monuments. *International Journal of Architectural Heritage*, 10(8), 1078–1100.

Pavoni, G., Giuliani, F., De Falco, A., Corsini, M., Ponchio, F., Callieri, M., and Cignoni, P. (2020). *Another Brick in the Wall: Improving the Assisted Semantic Segmentation of Masonry Walls*. GCH, pp. 43–51. http://vcg.isti.cnr.it/Publications/2020/PGD-CPCC20/GCH_2020_another_brick_in_the_wall_preprint.pdf.

Pavoni, G., Giuliani, F., De Falco, A., Corsini, M., Ponchio, F., Callieri, M., and Cignoni, P. (2022). On Assisting and Automatizing the Semantic Segmentation of Masonry Walls. *ACM Journal on Computing and Cultural Heritage (JOCCH)*, 15(2), 1–17.

Ravankhah, M., de Wit, R., Argyriou, A.V., Chliaoutakis, A., Revez, M.J., Birkmann, J., Žuvela-Aloise, M., Sarris, A., Tzigounaki, A., and Giapitsoglou, K. (2019). Integrated Assessment of Natural Hazards, Including Climate Change's Influences, for Cultural Heritage Sites: The Case of the Historic Centre of Rethymno in Greece. *International Journal of Disaster Risk Science*, 10(3), 343–361.

Romão, X., Paupério, E., and Pereira, N. (2016). A Framework for the Simplified Risk Analysis of Cultural Heritage Assets. *Journal of Cultural Heritage*, 20, 696–708.

Toy, S. (2006). *History of Fortification from 3000 BC to AD 1700 (No. 75)*. Barnsely: Pen and Sword.

Vandesande, A., and Van Balen, K. (2018). Preventive Conservation Applied to Built Heritage: A Working Definition and Influencing Factors. In: *Innovative Built Heritage Models*. London: CRC Press, pp. 63–72.

5 A Cultural Heritage Risk Index – The STORM Project Perspective

Mohammad Ravankhah, Maria João Revez, Rosmarie de Wit, Angelos Chliaoutakis, Athanasios V. Argyriou, Joern Birkmann, Apostolos Sarris, and Maja Žuvela-Aloise

Introduction

This chapter presents a systematic risk assessment and management methodology for cultural heritage sites, developed as part of the EU project 'Safeguarding Cultural Heritage through Technical and Organisational Resources Management' (STORM). The STORM project provides an innovative methodology, supporting tools, and services to assess and manage risks associated with natural hazards and climate change threats. The methodological framework was developed by a multidisciplinary team considering the existing approaches and methods in risk assessment and heritage conservation (e.g. Stovel, 1998; FEMA, 2005; D'Ayala et al., 2008; ISO 31010, 2009; UNESCO WHC et al., 2010; Birkmann and Welle, 2015; Michalski and Pedersoli, 2016) and particular characteristics of the cultural heritage site in the STORM project. Although the procedure portrays different steps sequentially for easier application, it is not a linear process. Conducting each step requires evaluation and revision of the previous ones, and also continuous monitoring and upgrading based on the projected hazard and vulnerability situation. The proposed procedure comprises the following major steps (Ravankhah et al., 2019a):

- Establishing the STORM context to determine the objectives and scope of the process
- Assessment of risks, including the hazard, exposure, and vulnerability analysis
- Treatment of risks to develop strategies for risk mitigation, preparedness, and recovery
- Implementation of the treatment strategies and monitoring the plan.

The proposed methodology of risk management was applied to five STORM pilot sites, the Historical Centre of Rethymno in Greece, the Mellor Heritage Project in the UK, the Roman Ruins of Tróia in Portugal, the Baths of Diocletian in Italy, and the ancient city of Ephesus in Turkey. Natural

DOI: 10.4324/9781003356479-6

hazards and threats affecting each pilot site were identified, analysed, and subsequently mapped in ArcGIS. After identifying and analysing the risk components, they were incorporated into a risk index to measure the risk levels. According to the STORM risk map concept, relative risk maps were generated to share a common understanding of the risks at the pilot sites among the risk management team, including site managers and stakeholders.

Site managers and key stakeholders were engaged as the main partners of the project consortium and actively contributed to the data collection and determining risk management strategies. In Tróia, for instance, the Tróia Resort, the Portuguese General Directorate of Cultural Heritage (DGPC), and the Grândola Municipality Civil Protection services were engaged in the project.

Following the risk assessment, risk treatment guidelines were provided to reduce the risks of sudden and slow-onset hazards. The three major pillars of the risk treatment framework consist of risk prevention and mitigation, including adaptation to climate change, risk preparedness, emergency response, and recovery. Accordingly, hazard and site-specific measures are proposed for implementing these strategies. The proposed measures include reducing hazards and threats, monitoring climate factors, setting up warning systems, reducing exposure, reducing material susceptibility, regular monitoring and maintenance of the sites, and emergency first aid. The proposed methods and outputs can facilitate decision-making by providing the spatial distribution of the significant hazards and a clear understanding of the vulnerability and risk within the pilot sites.

STORM Risk Assessment Procedure

'Risk assessment is the overall process of risk identification, risk analysis, and risk evaluation' (ISO 31000, 2009). STORM applies the risk index method for analysing risks to the pilot sites. 'Risk Index is a semi-quantitative measure of risk which is an estimate derived using a scoring approach using ordinal scales' (ISO 31010, 2009). Once the risk components are defined and measured, they are aggregated to create a composite risk index. The scores of the components are multiplied to rank different risks. The proposed risk index comprises the following risk components (see Figure 5.1):

- Hazard: sudden-onset hazards like storms, flooding, wildfires, and slow-onset hazards like change in freeze-thaw events, heat waves, and prolonged wet/dry periods were incorporated into the assessment procedure. Future alterations due to climate change, such as the projected change of precipitation and heatwaves, were also addressed in the assessment procedure.
- Exposure: movable and immovable heritage assets and their associated values are considered as elements at risk. Therefore, exposure assessment mainly focuses on analysing the heritage assets value within the pilot sites.

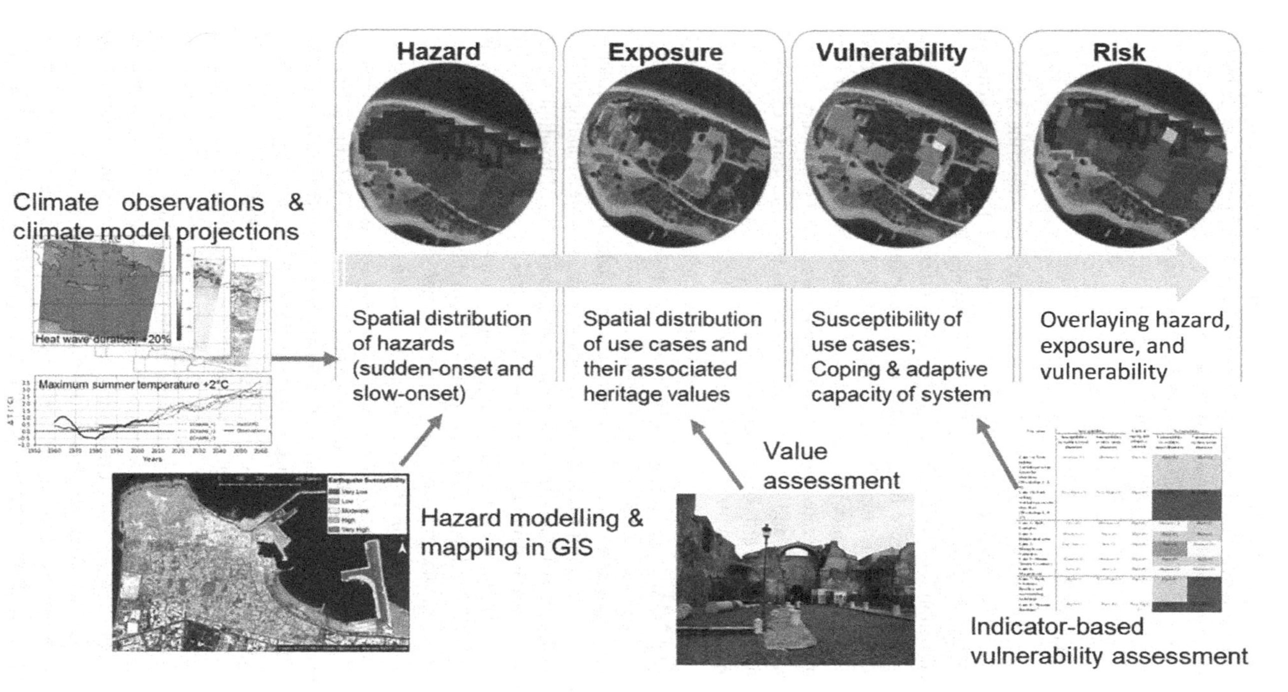

Figure 5.1 The concept of risk index and risk map for the pilot sites of the project.

Source: Author

- Vulnerability: an indicator-based vulnerability assessment method was developed to evaluate the susceptibility of the pilot sites to damage based on their structural and material characteristics. Furthermore, the adaptive and coping capacity of the conservation and management of the sites were incorporated into the vulnerability assessment.

Hazard Assessment

Natural hazards affecting the pilot sites were identified according to the 'STORM Classification of Hazards and Climate Change-related Events' (Ravankhah et al., 2019b). In addition to the common hazard classification, to adequately address the short- and long-term effects of natural hazards and threats on heritage sites, the hazards are further categorised according to the speed of onset. A semi-quantitative ranking (FEMA, 2004) was applied to analyse the potential hazards affecting the pilot sites and subsequently determine which hazards or threats need to be integrated further into the risk assessment procedure.

Hazard modelling and mapping were based on the availability of spatial data related to the hazards of interest. In this context, it was essential to specify the factors associated with the identified hazard via appropriate spatial analysis tools. The combination of hazard-related factors was implemented through GIS spatial analysis. For instance, factors associated with earthquake events such as the earthquake epicentres, proximity to active faults, and geological formations were combined to produce an earthquake hazard map. Another example is the tsunami hazard in the Tróia pilot site. Its corresponding hazard map was developed considering a 30 m resolution Digital Elevation Model (DEM) as an input data set. The modelling process was conducted through spatial analysis algorithms for a corresponding sea-wave height of 3–4 m infiltrating the mainland from a specific distance from the coastline (70–80 m).

In the climate hazard assessment, climate change effects relevant to cultural heritage are summarised. Based on the identification of atmospheric processes relevant to cultural heritage, an evaluation which was performed by cultural heritage experts, climate indices defined by the Expert Team on Climate Change Detection and Indices were assigned to these phenomena to aid a quantitative analysis. In this process, 'intense rainfall' is, for example, defined by the indices 'heavy precipitation days' and 'maximum one-day precipitation amount'. In total, 22 indices were chosen to define 14 climate-related hazards. To incorporate these results in the risk assessment procedure, the base hazard level (under current climatic conditions) and the relative climate change signal (rate of change relative to the model-based baseline) are determined. A final climate hazard assessment is made based on the relative signals obtained with the statistical and dynamical downscaling techniques and comparing the observed and modelled baseline situation to determine possible model biases. This assessment defines the hazard scale for each relevant climate index. The hazard level is assigned to one of five categories, ranging from very high to very low, based on which the climate hazards can be incorporated into the further risk assessment procedure.

Exposure Assessment

The term 'exposure' is widely agreed to correspond to elements-at-risk, such as people, resources, systems, and/or other valued assets whose location subjects them to potential adverse impacts from hazards (UNISDR, 2009). In the proposed framework, heritage assets and their tangible and intangible attributes are the elements at risk. Their exposure assessment is structured as follows:

- Description of heritage elements, including immovable and movable assets, within the site and its setting.
- Characterisation of the values of all heritage elements, based on a value category system (adapted from Worthing and Bond, 2008; Australia ICOMOS, 2013)
- Assignment of value levels, ranking the relative importance of the site elements.

The site value can be derived from institutional heritage listings, which state whether its importance is relevant at the local, regional, or national level. However, for site management in general and risk management in particular, institutional listings are insufficient for developing risk assessment and management. Thus, the chapter considers the value of the specific areas or assets at each pilot site individually rather than the overall value of the pilot site. This provides accurate data for further risk analysis and reduction strategies for each pilot site.

Vulnerability Assessment

An indicator-based method was applied to assess the pilot sites' vulnerability. Vulnerability analysis of the pilot sites and their assets is conducted through a structured questionnaire. The questionnaire is divided into two major sections, susceptibility analysis and coping and adaptive capacity analysis. The target group involves the pilot site managers, expert partners familiar with the sites, and local and national organisations responsible for site protection.

The susceptibility analysis is divided into four components that define the parameters and rank the elements, ultimately determining the susceptibility of a given heritage asset to suffer damage caused by sudden and slow-onset hazards. The components involve structure (load-bearing walls, foundations, roofs, and joints), structural materials (materials used in the load-bearing elements), (immovable) heritage interiors (e.g. decorative elements), and movable elements (e.g. collections and archives). Susceptibility to sudden-onset and slow-onset hazards are separately analysed to calculate their contribution to the overall vulnerability and risks adequately.

In addition to structural factors, the degree of capacity to mitigate, respond to, and recover from disasters contributes to the risk level. Coping capacities rely heavily on the institutional and management systems of heritage sites. In a broader context, it also relies on the regional and national bodies engaged in protecting cultural heritage from natural hazards. Coping and adaptive capacities

			Risk Index				
Areas	*Risk No.*	*Overall Risk statement*	*Hazard*	*Exposure*	*Vulnerability*		*Risk score*
					Susceptibility	*Lack of Coping and Adaptive Capacity*	
Sudden-onset disasters							
Hazard 1: Earthquakes							
RRT-la	Rl.la	*Potential structural cracks or total collapse of building components and archaeological elements at the Tróia site caused by an earthquake*	Medium (3)	Very High (5)	Medium (3)	High (4)	High (4)
RRT-lb	Rl.lb		Medium (3)	Very High (5)	Very High (5)	High (4)	High (4)
RRT-2	R1.2		Medium (3)	High (4)	Low (2)	High (4)	Medium (3)
RRT-3	R1.3		Medium (3)	High (4)	Medium (3)	High (4)	High (4)
RRT-4	R1.4		Medium (3)	Medium (3)	Very Low (1)	High (4)	Low (2)
RRT-5	R1.5		Medium (3)	High (4)	Medium (3)	High (4)	High (4)
RRT-6	R1.6		Medium (3)	High (4)	Low (2)	High (4)	Medium (3)
RRT-7	R1.7		Medium (3)	High (4)	High (4)	High (4)	High (4)
RRT-8	R1.8		Medium (3)	Medium (3)	High (4)	Very High (5)	High (4)

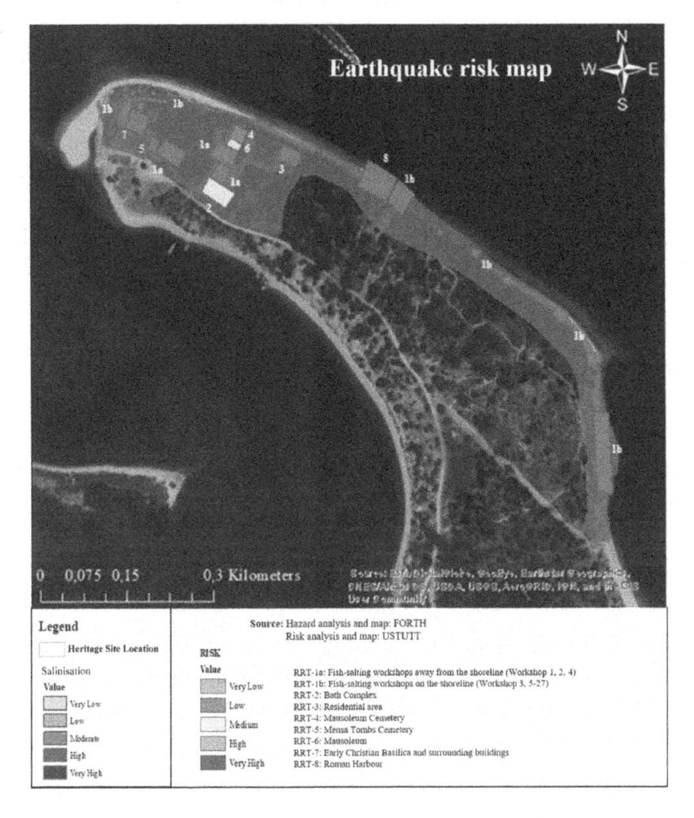

Figure 5.2 Developing an earthquake risk index (on the left) and an earthquake risk map (on the right) for Tróia.

Source: Author

are assessed by measuring a set of defined indicators, including multi-sectoral cooperation, risk awareness, information and communication systems, risk mitigation and preparedness, and monitoring and maintenance plans.

Risk Analysis and Evaluation

Following the risk components analysis, they were incorporated into the risk index to rate the level of the risks. The component scores were multiplied to rank different risks. For each hazard of interest, a risk statement representing the hazard's potential impacts on the site was defined. While considering the overall risk statement, the risk score for each site area (use cases) was separately calculated. Figure 5.2 illustrates an example of the earthquake risk index and map for the case of Tróia. The risk score may fall in one of the five equal-sized classes with class ranges of 1–1.8, 1.8–2.6, 2.6–3.4, 3.4–4.2, and 4.2–5. They are interpreted by numbers and colour codes, respectively, as Very low (1, dark green), Low (2, light green), Medium (3, yellow), High (4, orange), and Very high (5, red) level.

The ALARP (as low as reasonably practicable) principle was applied to evaluate and determine the risks that require further treatment strategies. Accordingly, very high and high risks (red and orange colours) were placed in the intolerable region, medium risks (yellow colour) in the tolerable region, and low and very low risks (green colours) in the acceptable region. This provides the further step of risk treatment by prioritising the risks and the heritage assets in the pilot sites while specifying which risks need to be treated.

Risk Reduction Strategies

Risk evaluation presents areas that require risk treatment. Risk treatment may consist of different options, including risk avoidance, removing risk sources, and changing the likelihood of hazards or consequences (AEMC, 2010). While considering these options, the proposed risk management framework provides cultural heritage sites with risk treatment strategies specific to sudden-onset and slow-onset hazards. For sudden-onset hazards, the framework comprises the following three major plans to address the pre, during, and post-disaster phases adequately:

- Risk prevention, mitigation, and climate change adaptation plan, including monitoring, maintenance, and conservation-restoration.
- Risk preparedness and emergency response plan.
- Recovery plan.

Several suggestions were compiled to aid the pilot sites in determining risk treatment strategies. The strategies and related measures are based on the risk components analysis and identification of the concern areas. Notably, heritage conservation and intervention principles play a key role in determining appropriate risk management strategies and must be adequately considered.

In the case of Tróia, structural measures include monitoring through real-time access to tide gauge data, photogrammetric documentation for damage assessment and digital conservation, and regular conservation works to consolidate and stabilise the surface finishes. Regular simulations, for example, earthquake and fire drills, were organised for staff along with civil protection services and other relevant local authorities, and a heritage and hazard information system by using a GIS database was established to enhance emergency response.

Once devised, the different options for treating each risk may be evaluated through a cost-effectiveness analysis (CEA). The STORM CEA methodology is a sequential process consisting of computing the initial investment and future costs of the different strategies and their effectiveness assessment via expert discussion, following a set of guidelines based on heritage conservation principles (Revez et al., 2019).

STORM Risk Assessment and Management Tool

To assist site managers and experts in assessing the level of hazards and risks in different areas of the site and determine site-specific strategies, a Risk Assessment and Management (RA&M) Tool was implemented according to the assessment methodology. The tool is developed as a web interface from the STORM main platform to help the domain experts and site managers. One of the key features of the tool is to provide risk assessment for multiple pilot sites, separate areas, and items for each site. The RA&M tool enables the site managers and experts to identify and analyse natural hazards affecting a heritage site, assess the area values of the site, analyse the vulnerability of the site, measure the level of risks in different areas of the site, and finally determine site-specific strategies to reduce the risk associated with each hazard.

The risk assessment tool and Web-GIS services provide spatial analysis of hazard, vulnerability, and risk to build situational awareness maps corresponding to different hazards and areas of each pilot site (Ravankhah et al., 2020). Web-GIS services can manage geospatial data to support risk assessment analysis, modelling, monitoring, and situational awareness processes. The corresponding Web-GIS service was implemented to support spatial (overlay) analysis of these data layers for understanding the scope, complexity, and severity of critical situations while identifying affected heritage structures and evaluating their potential damage. The respective Web-GIS map services are incorporated into the STORM platform to visualise critical situation events and associated hazards' severity as corresponding situational picture maps (Bicchierai et al., 2019). An example of the STORM platform utilising the situational awareness Web-GIS service is illustrated in Figure 5.3.

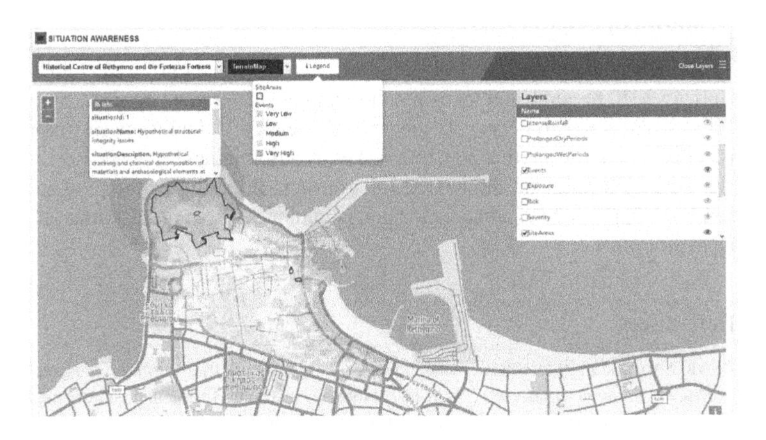

Figure 5.3 Situational awareness Web-GIS service of the STORM platform for the 'Historical Centre of Rethymno' pilot site.

Source: Author

Conclusion

The proposed risk assessment and management procedure provides a clear perception of the risk components, which is vital for determining hazard and site-specific risk reduction strategies. The risk assessment output further assists the decision-making process in understanding the elements that require treatment strategies and their respective levels. Accordingly, the strategies may include different options such as avoiding or reducing hazards, reducing structural susceptibility, and building coping and adaptive capacities.

The proposed method mainly targets cultural heritage sites, especially large-scale sites, to provide risk management for multiple hazards. The proposed methodological framework is applicable to different cultural heritage typologies. The criteria and indicators need to be adapted according to the site's material, structural and non-structural attributes, geological and socio-cultural characteristics, and the institutional and management systems.

The STORM RA&M tool provides a shared understanding of hazards and risks among multiple stakeholders engaged in protecting the pilot sites. The tool, combined with the Web-GIS services, facilitates the decision-making process in determining risk reduction measures for the different areas of each site and helps efficiently respond during an emergency. Overall, the proposed methodology addresses the different requirements of risk management for cultural heritage as defined by the multidisciplinary team of specialists and site managers. However, further research is needed to integrate the structural and non-structural components of vulnerability for different typologies of cultural heritage at risk while addressing the spatial and temporal dynamics of vulnerability.

Acknowledgements

This article is based on the STORM project, funded by the European Union's Horizon 2020 research and innovation programme under grant agreement no. 700191. The article reflects only the authors' views, and the European Union is not liable for any use that may be made of the information contained herein. We would like to thank the STORM consortium, in particular the pilot site staff, for supporting the consultation and communication in the risk management procedure.

References

AEMC (Australian Emergency Management Committee). (2010). *National Emergency Risk Assessment Guidelines (NERAG)*. Hobart, Australia: Tasmanian State Emergency Service. www.preventionweb.net/publications/view/41033.

Australia ICOMOS (International Council on Monuments and Sites). (2013). *The Burra Charter: The Australia ICOMOS Charter for Places of Cultural Significance*. Burwood,

Australia. http://australia.icomos.org/wp-content/uploads/The-Burra-Charter-2013-Adopted-31.10.2013.pdf.

Bicchierai, I., Gugliandolo, E., Williamson, R., Ravankhah, M., Chliaoutakis, A., Argyriou, N., Sarris, A., and Argyriou, L. (2019). Decision Making for Risk Mitigation Based on Collaborative Services and Tools. In: Resta, V., Utkin, A.B., Neto, F.M., and Patrikakis, C.Z. (Eds.), *Cultural Heritage Resilience Against Climate Change and Natural Hazards Methodologies, Procedures, Technologies and Policy Improvements Achieved by Horizon 2020–700191 STORM Project*. Pisa: Pisa University Press, pp. 55–88.

Birkmann, J., and Welle, T. (2015). Assessing the Risk of Loss and Damage: Exposure, Vulnerability and Risk to Climate-Related Hazards for Different Country Classifications. *International Journal of Global Warming*, 8(2), 191–212. https://doi.org/10.1504/IJGW.2015.071963.

D'Ayala, D.F., Carriero, A, Sabbadini, F., Fanciullacci, D., Ozelik, P., Akdogan, M., and Kaya, Y. (2008, October 12–17). *Seismic Vulnerability and Risk Assessment of Cultural Heritage Buildings in Istanbul*. Conference Session. The 14th World Conference on Earthquake Engineering, Beijing, China. www.iitk.ac.in/nicee/wcee/article/14_S11-079.PDF.

FEMA (Federal Emergency Management Agency). (2004). *Using HAZUS-MH for Risk Assessment: How-to-Guide*. Washington, DC: Federal Emergency Management Agency. http://mitigationclearinghouse.nibs.org/content/fema-433-using-hazus-mh-risk-assessment-how-%C2%A0guide-2004.

FEMA (Federal Emergency Management Agency). (2005). *Integrating Historic Property and Cultural Resource Considerations into Hazard Mitigation Planning: How-to Guide (FEMA 386–9)*. Washington, DC. http://wyohomelandsecurity.state.wy.us/grants/hmpg/Integrating_Historic_Property_Cultural_Resource_Considerations_into_hmplanning.pdf.

ISO (International Organization for Standardization) 31000. (2009). *Risk Management – Principles and Guidelines: BS ISO 31000:2009 = Management du risque – principes et lignes directrices*. Geneva: International Organization for Standardization.

ISO (International Organization for Standardization) 31010. (2009). *IEC 31010:2009: Risk Management – Risk Assessment Techniques*. Geneva: International Organization for Standardization.

Michalski, S., and Pedersoli, J.L. (2016). *The ABC Method: A Risk Management Approach to the Preservation of Cultural Heritage= La méthode ABC pour appliquer la gestion des risques à la préservation des biens culturels*. Canadian Conservation Institute and ICCROM. www.iccrom.org/publication/abc-method-risk-management-approach-preservation-cultural-heritage.

Ravankhah, M., Chliaoutakis, A., Revez, M.J., de Wit, R., Argyriou, A.V., Anwar, A., Heely, J., Birkmann, J., Sarris, A., and Žuvela-Aloise, M. (2020). A Multi-Hazard Platform for Cultural Heritage at Risk: The STORM Risk Assessment and Management Tool. *IOP Conference Series: Materials Science and Engineering*, 949(1), 012111.

Ravankhah, M., de Wit, R., Argyriou, A.V., Chliaoutakis, A., Revez, M.J., Birkmann, J., Žuvela-Aloise, M., Sarris, A., Tzigounaki, A., and Giapitsoglou, K. (2019b). Integrated Assessment of Natural Hazards, Including Climate Change's Influences, for Cultural Heritage Sites: The Case of the Historic Centre of Rethymno in Greece. *International Journal of Disaster Risk Science*, 10, 343. https://doi.org/10.1007/s13753-019-00235-z.

Ravankhah, M., de Wit, R., Revez, M.J., Chliaoutakis, A., Argyriou, A.V., Vaz Pinto, I., Brum, P., Žuvela-Aloise, M., Sarris, A., Birkmann, J., Anwar, A., and Heeley, J. (2019a). Risk Assessment and Risk Management for the Protection of Cultural Heritage. In: Resta, V., Utkin, A.B., Neto, F.M., and Patrikakis, C.Z. (Eds.), *Cultural Heritage Resilience Against Climate Change and Natural Hazards Methodologies, Procedures, Technologies and Policy Improvements Achieved by Horizon 2020–700191 STORM Project*. Pisa: Pisa University Press, pp. 55–88.

Revez, M.J., Coghi, P., Delgado Rodrigues, J., and Vaz Pinto, I. (2019). Analysing the Cost-Effectiveness of Heritage Conservation Interventions: A Methodological Proposal Within Project STORM. *International Journal of Architectural Heritage*. https://doi.org/10.1080/15583058.2019.1665141.

Stovel, H. (1998). *Risk preparedness: A Management Manual for World Cultural Heritage*. Rome: ICCROM. www.iccrom.org/ifrcdn/pdf/ICCROM_17_RiskPreparedness_en.pdf.

UNESCO WHC, ICCROM, ICOMOS, and IUCN. (2010). *Managing Disaster Risks for World Heritage*. Paris: World Heritage Resource Manual, UNESCO. http://whc.unesco.org/uploads/activities/documents/activity-630-1.pdf.

UNISDR (United Nations Office for Disaster Risk Reduction). (2009). *Terminology on Disaster Risk Reduction*. Geneva: UNISDR. www.unisdr.org/files/7817_UNISDRTerminologyEnglish.pdf.

Worthing, D., and Bond, S. (2008). *Managing Built Heritage: The Role of Cultural Significance*. Oxford: Blackwell Publishing.

Part 2

DRM Plan Implementation – Workshops/Community Engagement/Traditional Knowledge

6 Mapping Risk for Cultural Heritage

A Project on Archaeological Decorative Elements in Mexico

Dulce María Grimaldi and Mónica Vargas

Introduction

Mexico, a highly populated country, has numerous heritages linked with religious beliefs, social practices, and the economic income through tourism (see Figure 6.1). Therefore, the National Economic and Development Plan *(2019–2024)* developed by INAH includes risk prevention for heritage among its objectives.

Through time, the Mexican cultural heritage has suffered from various natural and anthropogenic hazards like seismic and volcanic activity in the central and southern regions, instability of mountain slopes, cyclones, heavy rain precipitation, tsunami and flooding in the coastal areas, fire incidents linked to extensive gas pipelines, and massive crowding on special occasions, among others. The recent disasters caused by the earthquakes in 2017, 2018, and 2020 seriously damaged the built cultural heritage and traditional practices, disrupting people's life. Consequently, these events emphasise that prevention is more accessible, cost-effective, and less emotionally wrecking than response and recovery.

On the other hand, the written documents and protocols concerning risk management produced before such disasters occurred were insufficient to prevent their impact on the built cultural heritage. The documents were not disseminated along with an education programme instrumented for the cultural heritage personnel and the society. Therefore, during earthquakes, lack of familiarity with the preventive actions favoured slow response. Hence, it is crucial to generate documents, implement action for dissemination and capacity building, and constantly update the information. Furthermore, awareness must be instilled about the recurrence of these disasters.

The proposed initiative is simple and encourages action to collaborate for risk prevention and quick response planning. Risk assessment through mapping was proposed and launched by the personnel of the Mexican National Agency for Cultural Heritage in Mexico (INAH) that works on a regional conservation project (*Proyecto de Conservación de Bienes Culturales Muebles*

DOI: 10.4324/9781003356479-8

Figure 6.1 Collapsed shelter over the wall painting and stucco reliefs and pavements due to Hurricane Grace, at the Archaeological Site of Tajín, Mexico.

Source: Photograph by Fidel Ugarte

Asociados a Inmuebles Arqueológicos en la Región Centro-Sur-Golfo de México- PCSG). Decorative elements of built heritage from archaeological sites in the central and southern regions were chosen as the project's scope. This includes exterior wall paintings, sculptured reliefs, renderings, and pavements that belong to the Mesoamerican civilisations before the Spanish conquest. The selected region has a high concentration of built cultural heritage, coinciding with the heritage area damaged by earthquakes and is one of the country's low-economic areas. Therefore, heritage also plays a vital role in society's well-being.

Hence, risk mapping was added to the current work undertaken for 62 archaeological sites in this region, including four World Heritage Sites: Tajin, Monte Albán, Xochicalco, and Teotihuacan. The project attempts to address four major questions:

1. Which decorative elements are to be considered?
2. Where are they located?
3. What is the condition of the art?
4. Are those decorative elements at risk?

Methodology

The methodology for risk mapping was developed during the ITC in 2016 at Ritsumeikan University and later tropicalised to suit the Mexican requirements. The mapping takes place at different stages and includes mapping of specific cases to obtain risk and priority maps of the whole region, consequently assisting in defining preventive measures and response strategies.

The documentation on local hazards, both natural and manufactured, is prepared, and those applicable at each archaeological site are selected. In addition, documentation for the decorative elements is prepared during the site visits, where each element's cultural and physical characteristics and images are gathered on a file that serves as an individual identity chart. Finally, the decorative elements are located on a map.

Since there is a vast expanse of decorative elements, only valuable and relevant ones must be selected to simplify the work. The proper selection of the relevant decorative elements was the foremost challenge encountered in the methodology. As overall values are considered, unique universal values as a guide were used along with the commonly shared opinion about relevance from the society. Therefore, single selection criteria for decorative elements are not viable. The criteria may be tweaked according to each archaeological site.,

Subsequently, vulnerability, possible impact, and previous events are analysed to generate a risk map. Ultimately, priority maps and preventive measures to reduce the vulnerability and impact of the hazard are proposed.

The methodology implementation undergoes different stages, beginning with specific cases and concluding with regional analysis. The implementation stages include documentation on-site, data analysis in the laboratory, workshops to promote the involvement of local stakeholders, data collation, and concluding with regional assessment and mapping.

The first stage includes documenting the existing decorative elements and evaluating their relevance on a three-level score. Furthermore, information about vulnerability, possible impact and previous events is gathered in consideration with the baseline information about regional and local hazards. In addition, the vulnerability of the decorative elements towards specific hazards is also mapped.

Training in primary information collection is necessary for collaborative work at this implementation stage. The training introduces risk management concepts to formulate a standard glossary and understanding of the different stages of prevention, response, and recovery, along with developing a homogeneous way of filling up the formats.

In the second stage, the collected information is analysed in the laboratory. The collected data is further classified and added to a database. For instance, vulnerability is classified according to location, the sensitivity of constituent materials, lack or insufficient maintenance, and previous interventions and infrastructure. Consequently, the frequently shared vulnerability of the decorative elements at each site is identified, and preventive control measures for those vulnerabilities are proposed, such as maintenance.

Simultaneously, the severity of the possible impact is rated on four levels, high, medium, low, or almost non-existent, and through analysis, the hazard with severe effect is identified. However, it might not be similar to one with a negative impact on most decorative elements. Previous events are recorded by number and severity of the damage caused. At the same time, the dates of occurrence provide a general idea about the frequency of the events.

The resulting risk maps consist of numerous information. The first one gathers complete site information about the vulnerability of each decorated element towards all considered hazards, marked on colour squares (see Figure 6.2). Separate maps show the vulnerability of the decorative elements towards a specific hazard to enable easier reading. Finally, the priority map helps visualise the areas with the highest rate of risk, which concentrates on more vulnerable decorative elements, endangered by a large number of hazards, potentially suffering the worst impact (see Figure 6.2).

The second stage also includes the analysis of collected data, both written and graphic, which helps to formulate an elaborate list of recommended preventive measures.

Stakeholders from the cultural institution, the government, and society are involved in the third stage of implementation. Draft material in the form of maps and recommended preventive measures are shared with the stakeholders to encourage their involvement and offer their input with accurate information. The local communities can provide valuable information as they are familiar with the tradition and risks at the site, having dealt with it for a long. Currently, such involvement is restricted to the site authorities. Thus, future workshops with a broader group of stakeholders could yield better results.

At this stage, it is realised that the awareness of the risk is closely connected to the stakeholders' involvement. Hence, this initiative's graphic and written documents concentrate on information about the risk to the cultural heritage. Risk management measures could prove ineffective if stakeholders are unwilling to consider this information. Thus, it is vital for the National Agency for Cultural Heritage in Mexico (INAH) to consider this initiative in the working strategies for risk management and provide an adequate budget.

Finally, at the data collation and regional assessment stage, the information gathered from every site is collated on a database. The database forms the baseline information for creating geographic information system maps that can aid regional assessment. Therefore, the current risk maps for cultural heritage in Mexico can become an accurate indicator for the decorative elements at archaeological sites in future. Furthermore, the mapping must be updated periodically with ongoing disasters, which will help in preventing

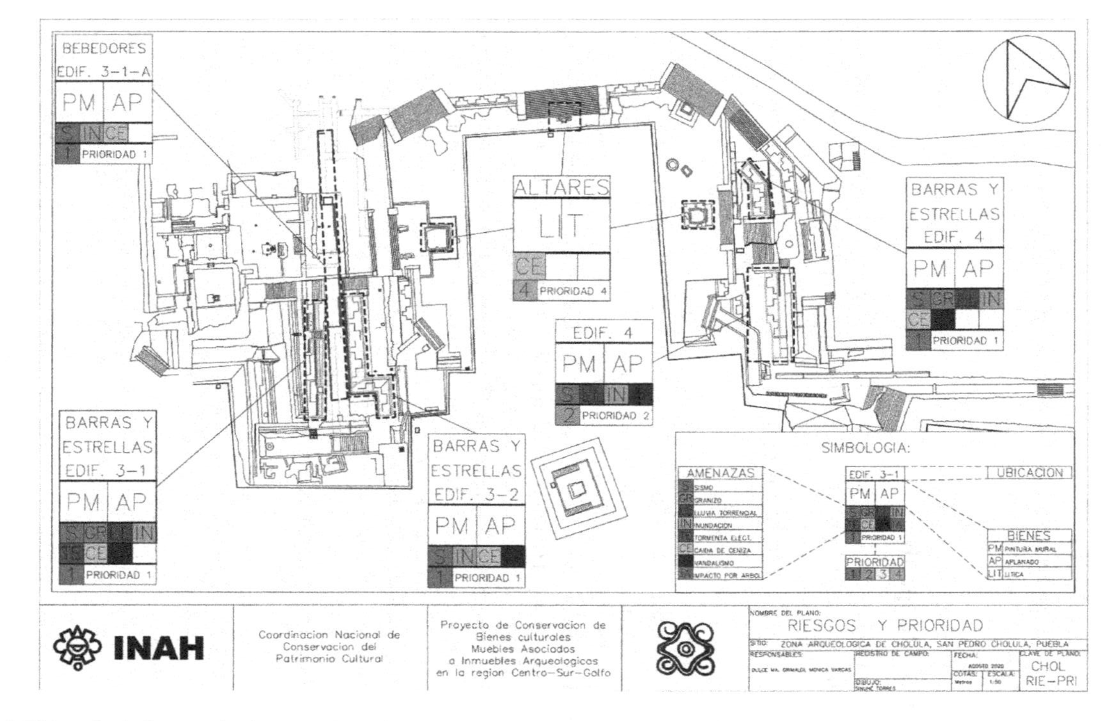

Figure 6.2 Risk and priority map for the case study of the decorative elements at the Archaeological Site of Cholula, Mexico.

Source: Author

and planning quick response activities at the sites. The risk maps will assist in answering the following questions:

1. How many decorated elements are at risk due to a specific hazard, and where are they located?
2. Which areas require prompt response after a specific disaster?
3. What assistance can be offered to people to recover from cultural heritage loss after a disaster?

Future Perspective

The project began in 2018, and so far, risk assessment and mapping have become an integral part of every conservation approach for decorative elements in the area encompassed by the ongoing project. The methodology has been adopted to Mexican requirements, while capacity building and awareness among the cultural heritage professionals have led to prioritising risk management for cultural heritage.

Until now, the implementation for eight case studies has been completed along with two regional assessments. Due to COVID-19 restrictions, the project implementation was slow-paced. However, the project documents the cultural heritage considered for protection, its relevance, and the kind of hazards and vulnerabilities that usually occur. Furthermore, the information has enabled identifying the cultural heritage under higher risk and determining probable mitigation strategies. Priorities are established for each site and its regional scope. However, the earthquake in 2020 proved that significant work remains to make risk assessment and mapping an effective risk management tool.

Society must collaboratively use mobile devices to gather quick information and generate real-time shared digital map for prompt response. Proper education and dissemination programmes among the cultural heritage professionals and society are missing. Thus, it is important to include numerous activities through workshops and graphic reading material to strengthen the strategy of involving multiple stakeholders. Stakeholders from the society, national cultural agency personnel, and government need to be smartly approached and encouraged to participate and support the initiative, which until now has been a challenging task in the project.

Since this is an enormous initiative, it must be simple yet encourage conservators, cultural heritage professionals, and local communities to work collaboratively. While the project focuses on working on decorative elements at archaeological sites, the purpose is to provide a working model for all types of decorative elements, including the historical. In addition, it enables the assessment for all elements of the cultural heritage, decorative, built, tangible, and intangible. Therefore, the methodology and working stages must be tested for a broader scope.

This is a beginning step, among others, towards creating a risk management plan for the built cultural heritage of Mexico to prevent it from being stalled under the list of priorities.

Acknowledgement

We would like to thank the authorities of the Department for Conservation of Decorative Elements from the National Agency for Cultural Heritage in Mexico (*Coordinación Nacional de Conservación del Patrimonio Cultural del INAH*) for their support and acceptance to the inclusion of this topic to the ongoing regional conservation project (Maria del Carmen Castro, Salvador Guiliem, Cristina Ruiz). As well, we would like to recognise the enthusiastic collaboration of the working team from the PCSG project (Miriam Segura, Omar Torres, Alfonso Osorio, Sinuhé Torres, and Claudia López). Finally, we thank the commentators and moderators of the 'Workshop on Good Practices for the Disaster Risk Management of Cultural Heritage 2020' for their valuable input to this initiative, along with the inspiration promoted by the presentation of other good practices from the rest of the colleagues at this forum.

7 George Town World Heritage City's Disaster Risk Management

Ming Chee Ang

Introduction

In 2008, George Town in Penang, Malaysia, was inscribed as the World Heritage Site of 'Historic Cities of the Straits of Malacca, Melaka and George'. The World Heritage Committee inscribed this serial nomination based on the criteria ii, iii, and iv.[1]

In 2010, the George Town World Heritage Incorporated (GTWHI) was established by the State Government of Penang as the site manager for the George Town UNESCO World Heritage Site to safeguard and conserve the outstanding universal value of the World Heritage Site.[2] The GTWHI team is chaired by the Chief Minister of the State Government of Penang, guided by the Board of Directors, and consists of 36 full-time professionals, all dedicated to the conservation and safeguarding of the tangible and intangible heritage.[3]

Since 2017, GTWHI was tasked with implementing the UNESCO pilot case studies for the project 'Capacity Building for Disaster Risk Reduction of Heritage Cities in Southeast Asia and Small Island Developing States in the Pacific'. The project was part of UNESCO and Malaysia's commitment to implementing the Sendai Framework for Disaster Risk Reduction (2015–2030). The framework was adopted during the Third World Conference on Disaster Risk Reduction held in Sendai, Japan, in March 2015. The Sendai Framework reviews the former Hyogo Framework by aligning itself with UNESCO's sustainable development goals and developing indicators for efficient monitoring. It also highlights the need for collaboration and integration. Heritage should be included in national and local disaster risk management plans. Inversely, disaster risk must be addressed in the overall heritage management plans of the World Heritage Sites.

The UNESCO pilot project on 'Capacity Building for Disaster Risk Reduction of Heritage Cities in Southeast Asia and Small Island Developing States in the Pacific' was funded by the Malaysian Government under the Malaysian Funds-In-Trust. The pilot sites for this project are (i) Melaka and George Town, Historic Cities of the Straits of Malacca of Malaysia, (ii) Kota

DOI: 10.4324/9781003356479-9

Lama in Semarang of Indonesia, and (iii) the Levuka Historical Port Town of Fiji. The project's main objective is to enhance the capacity of stakeholders involved in the Disaster Risk Reduction of heritage cities in Southeast Asia and the Pacific. It also aims to empower communities living in the heritage cities through improved management plans, including Disaster Risk Reduction strategies. For George Town, this project is a game changer in disaster risk reduction on cultural heritage, with meaningful community involvement, instead of a mere World Heritage Site project.

Disaster Risk Management for Cultural Heritage

The inception workshop for the UNESCO pilot project on 'Capacity Building for Disaster Risk Reduction of Heritage Cities in Southeast Asia and Small Island Developing States in the Pacific' was hosted on April 5–6, 2017 in George Town, Penang. This workshop aims to identify the efforts, challenges, and gaps in the Disaster Risk Reduction Programme of the participating cities. Important concepts such as hazards, exposure, and vulnerability were introduced to the main stakeholders of the World Heritage Site.

Fire and water were identified during the workshop as the two major hazards within the George Town UNESCO World Heritage Site. Heritage buildings such as the shophouses are vulnerable to fire due to their materials, proximity to each other, and narrow streets that limit accessibility for fire trucks. George Town also suffers from floods due to excessive rain and high tides. Considering the urgency of these risks, GTWHI pledges prompt, efficient, and collaborative work with the local authority and residents to raise awareness and reduce the risks.

The disaster risk reduction management plan prepared during the UNESCO Chair Programme on Cultural Heritage and Risk Management at the International Training Course on Disaster Risk Management of Cultural Heritage in September 2017 was used when George Town suffered from one of the worst floods in the state's history on November 4–5, 2017. Working under the concept of building back better, George Town continues to evolve and improve its heritage resilience to disasters post the November 2017 flood.

The Disaster Risk Reduction Strategy of George Town was prepared at the national workshop on 'Harmonizing Coordination to Implement Disaster Risk Reduction Strategy' in George Town in March 2018. A total of 64 participants representing the central, state, and local-level governments, community associations and groups, civil defence, first responders, and site managers participated. The attendance of important government representatives represented the shared commitment of the government and public sectors in enhancing the disaster risk reduction agenda in the public services of George Town and Penang. The event was covered by more than 13 printed and online media. The wide coverage in newspapers and online media built strong branding and sent a positive message on the disaster risk recovery and management strategy in George Town.

During the March 2018 National Workshop, ten action plans were developed by the George Town community. GTWHI was tasked to strategise the priority and implementation of these action plans along with the community members. Each action plan is elaborated according to the planning context, implementing strategies, and the output achieved.

Implementation of Action Plans

In any heritage programme, the genuine involvement of the main stakeholders is an important factor in its success. GTWHI utilised a social network of local agencies and community and invited them to join the action plans. Key players and first responders during disaster management, such as the firefighting department, self-defence department, local community leaders, and clan councils who own a significant number of heritage properties within the World Heritage Site, were mobilised. Their inputs are organised into the following action plan for George Town.

Action Plan 1: Enhancement of Cultural Heritage Elements in Disaster Risk Management

It is important to include cultural heritage elements and professionals in the planning and execution of disaster risk management. Since 2017, lobbies were created, and documentation was prepared to include cultural heritage elements and perfectives in the local and national disaster risk management legislation. This includes suggesting that the National Disaster Management Agency incorporate cultural heritage elements in the National Security Council's Directive No. 20. Adding the National Heritage Commissioner, State Heritage Commissioner, local council Heritage Department, and World Heritage Site Managers to the Disaster Management Committee was recommended. Furthermore, adding the Department of National Heritage from the Federal Government, the Department of Heritage Conservation from the local municipal, and the World Heritage Site Manager Office as the site management agency into the Disaster Management Responding Agency was also recommended. In addition, a suggestion was made to establish the 'Heritage Materials Management Centre' for the handling, packing, and storing of salvaged heritage objects.

To improve the legislation within the World Heritage Site, efforts were undertaken to revise the George Town Special Area Plan with a new chapter on disaster risk management to be incorporated. There is also ongoing lobbying to prepare the 'Fire Safety Guidelines for Heritage Buildings' with the Malaysian Institute of Architects.

Action Plan 2: Community-Based Capacity Building

The methodologies for George Town's community-based pilot cases were developed based on the disaster risk management for cultural heritage by the Institute of Disaster Mitigation for Urban Cultural Heritage. In George Town, training was conducted in a few workshops, each lasting two hours. Long-duration training, often four to eight weeks, was unsuitable. Since many local communities have day jobs, engaging in long-term training will be difficult.

The first workshop of the 'Community-Based Disaster Risk Reduction Coordinating Workshop for George Town UNESCO World Heritage Site' was conducted on September 4, 2018. This workshop introduced basic concepts of hazard, vulnerability, risk, disaster, emergencies, and attributes with examples. A site mapping and imagination game were also conducted. During the second workshop on November 2, 2018, the participants were invited to share a disaster scenario and mitigation for human life and cultural heritage. The participants also discussed their emergency response strategies. By the third workshop held on January 8, 2019, the participants began detailing the risk analysis, identifying potential stakeholders, and defining their roles before, during, and after the disaster. The participating cases were presented during the fourth workshop held on March 22, 2019, including the Standard Operation Plan of the cases before, during, and after a disaster. During the Experience Sharing Workshop, held on June 21, 2019, all participants were provided technical assistance prior to their presentations.[4]

Action Plan 3: Backup Copies of Cultural Heritage Inventory List

During the community workshops, GTWHI identified the needs and opportunities in safe keeping backup copies of heritage properties blueprint and the inventory list of cultural heritage artefacts for the local heritage stakeholders. Through such understanding, a few major clan houses in George Town agreed to share their cultural heritage inventory list with GTWHI. Plans were also made to collect important documents, such as measured drawings, dilapidation reports, cultural objects inventory lists, and other documentation from heritage building owners in the George Town UNESCO World Heritage Site. Such documents are now digitised and their backup has been taken at the GTWHI database.

Action Plan 4: Geographic Information System in George Town Disaster Risk Reduction

Fire is identified as a major disaster risk threatening George Town, and efforts are made to map and identify the locations of fire hydrants in the World Heritage Site. The arrival time and route of the first responder are studied to ensure that fire brigades can arrive and provide help timely. GTWHI has enhanced the collaboration with the Fire and Rescue Department of Malaysia to update the fire hydrant locations in George Town and ensure that all buildings in the World Heritage Site are within reach of a functioning fire hydrant.

Action Plan 5: Effective Monitoring With Technology

The City Council of Penang Island has installed nearly 952 units of closed-circuit television cameras equipped with high-resolution technology at significant spots to monitor vehicle traffic, flash floods, and public safety. Approximately 30 flood sensor devices are installed in several flood-prone areas to help the council monitor and track the flash flood occurrences on the island and timely mobilise the authorities. Water levels are marked in green when below 75% of the height of the drain, yellow when it reaches the alert level, orange as warning level when the water level has reached 75% of the height of the drain, and red as the danger level when the road surface is flooded with the water level rising to 250 mm. Warning alerts are channelled to the City Council of Penang Island's 24-hour hotline through SMS, alerting the City Council's Squad *Pantas* to immediately activate their Standard Operating Procedure for preventive action in disasters.

Action Plan 6: Public Awareness Campaigns

Since 2017, campaigns have been conducted by GTWHI to raise public awareness and enhance public capacity to react calmly during a disaster. In addition, AED training has been conducted in George Town since August 2018. The first AED unit was also prepared for public use at the GTWHI premises. In December 2019 and January 2022, the AED machine resuscitated casualties during an acute cardiac arrest incident and a drowning case.

To expand the awareness campaign beyond the World Heritage Site, GTWHI printed the Disaster Risk Management on Cultural Heritage posters in four languages, which were enthusiastically received by the local community and government agencies. The design and content were based on the feedback received from the community during the first session of the community workshop. Raising awareness through posters was the best approach for the local communities as it included the local context and was easier to understand.

Action Plan 7: Emergency Response Teams of George Town

Since 2017, GTWHI has set up an Emergency Response Team. The GTWHI staff members and the community were invited to join the firefighting theory and practical sessions conducted by the Fire and Rescue Department of Malaysia. Emergency Response Team Training Sessions were conducted on May 4–5, 2019, by the Head of the Beach Street Fire and Rescue Department. The session provided the safety briefing and introduction and explained the role of the Emergency Response Team. The second training day focused on practical training, such as basic rescue techniques and fire hydrant drills at Seh Tek Tong Cheah Kongsi.

All the participants underwent the Emergency Response Team training exam and received one kilogram of dry powder fire extinguisher and a certificate of completion on the second day. The establishment of the response team prepared GTWHI to react promptly and calmly during an emergency and created a strong bond with the firefighters and rescue teams from George Town. They have supported each other in many trainings, site inspections, community work, and other developments, including establishing the Lebuh Pantai Fire and Rescue Station Gallery in 2019.

Action Plan 8: Community-Based Fire Preparedness and Response Strategy

As there are more than 5,300 buildings within George Town, UNESCO World Heritage City, it is important to proactively prepare the site to reduce the risk of fire incidents. The GTWHI has targeted to provide at least one smoke detector and a fire extinguisher for every building within the George Town World Heritage Site.

The first phase of this plan began in 2019. Notifications were published in major newspapers and online platforms, inviting qualified community members to participate in the programme on January 30, 2019. Premises owners and residents within George Town UNESCO World Heritage Site could submit their applications online. Community members with no Internet access could visit the World Heritage Site Manager's office for assistance in submitting their applications. Under this project phase, 50 applicants were selected. Selected participants underwent firefighting training, jointly organised by the GTWHI and the Fire and Rescue Department of Malaysia.

A public ceremony was held to inaugurate the first batch of Community-Based Fire Responders in the presence of the media and attended by senior representatives from the State Government of Penang. The GTWHI conducted house visits to all the participating units to review their basic fire protection measures and provide advice for improvements. After selecting 50 applicants, two training sessions were arranged for local residents and participants in George Town UNESCO World Heritage Site from April to May 2019.

From July 6, 2020, fire extinguisher maintenance was conducted for the first phase of Community-Based Fire Responders. The GTWHI team visited 50 participant houses to collect their fire extinguisher, check the pressure, condition of the ABC powder, and renew the Fire and Rescue Department's certificate displayed on the fire extinguisher. During the house visit, the GTWHI team also discussed with the participants to understand their condition during the COVID-19 pandemic. Most of the participants responded that they could overcome the difficult period, but it is evident that their businesses have suffered due to the pandemic.

Action Plan 9: First Aid for Cultural Heritage and Damage Assessment

The GTWHI team referred to ICCROM's 'How to Undertake Integrated Post-Event Damage and Risk Assessment' document for executing first aid for cultural heritage and damage assessment. The Standard Operating Procedure for damage assessment was first developed in December 2017 to access the landslide at Penang Hill. In July 2019, the form was used again when the GTWHI team assessed the Bukit Mertajam Xuan Tian Temple, built in 1886 and had caught fire on July 3, 2019, at 8:30 pm. The team also visited the community affected by the fire at Lebuh Carnarvon and conducted cultural heritage rescue and assessment missions in February 2020.

Collaborative digital work enabled producing assessment reports within 24 hours after reviewing any damaged site. Safety preparations for the GTWHI staff, such as helmets, gloves, shoes, ventilation masks, protective eye goggles, and reflective safety vests, were compulsory before entering any damaged site. Permission from the Fire and Rescue Department of Malaysia is mandatory before entering the site, and the visit is completed in under 15 mins. A camera is installed on one of the team members' helmets to record the visit, and drones are also used sometimes to record post-disaster videos and pictures. These documentations and follow-up discussions with the affected community offer a significant learning opportunity for the GTWHI to empathise, listen, and work with the community to rebuild their properties.

Community-Based Disaster Risk Management

George Town is a UNESCO World Heritage Site with 5,013 buildings, and about 78% of these buildings are heritage buildings. The site has 10,000 residents, with an average of some 100,000 daytime and night-time users daily. Through this Disaster Risk Management for Cultural Heritage in George Town, efforts were made to empower the local residents and users of the site. More importantly, capacity building on resiliency is crucial to ensure that people can embrace modernisation and community progression through disaster risk reduction at all levels of society.

Cultural heritage is a shared asset and safeguarding it from disasters must be a collective effort. Since the inception workshop in April 2017, GTWHI has learnt that people are the most important resource for a successful disaster risk management programme. Communication between heritage building owners and local rescue departments is important to bridge the information gap. Local leaders are key to mobilising the buy-in of the local people on the knowledge and approaches. The willingness of professionals to contribute their knowledge and time to the project will help GTWHI to work with more local residents, building owners, and stakeholders, making George Town a resilient city.

Above all, disaster risk management is a long-term commitment and strategies for its implementation must consider sustainability. Often, the focus is on financial allocations as a key success factor for any project. However, George Town's approach focuses on the local community's involvement and collectiveness. It demonstrates that framing disaster risk reduction as a tool for safeguarding one's life and property is more efficient. Thus, it is important to introduce the concepts and methods of disaster risk management in the local language for better understanding.

The lessons from George Town could inspire more heritage cities to invest time and resources in building capacity within the community to make their city safer and more resilient to risks.

Acknowledgement

The author would like to thank the GTWHI team and the George Town community for supporting the George Town cultural heritage disaster risk reduction efforts since 2017.

Notes

1 UNESCO, https://whc.unesco.org/en/list/1223/.
2 The Melaka Heritage City is managed by the Melaka World Heritage Office.
3 For more information on George Town World Heritage Incorporated, please visit: https://gtwhi.com.my/.
4 The cases that were presented included from George Town World Heritage incorporated by Ms Shereen Loh Phaik See, Boon San Tong Khoo Kongsi by Mr Khoo Teng Khoon, Seh Tek Tong Cheah Kongsi by Mr Peter Cheah Swee Huat, Sia Boey by Ms Virajitha Chimalapati, Toh Aka Lane by Mr Teng Wei Yee, and Lim Jetty by Mr Patrick Lim.

8 Historic Water Cisterns – An Effective Fire Preventive System

Elena Mamani

Introduction

This chapter is based on a presentation that was given during the online workshop titled "Good Practices for Disaster Risk Management of Cultural Heritage", which was organized by the Institute of Disaster Mitigation for Urban Cultural Heritage, Ritsumeikan University (RDMUCH) and the International Centre for the Study of the Preservation and Restoration of Cultural Property (ICCROM) in October 2020. The presentation focuses on the concept of revitalizing traditional rainwater collection systems and utilizing them as a preventive firefighting system to safeguard historical buildings. The idea originated during a case study conducted at the International Training Course on Disaster Risk Management in Kyoto, Japan, in 2014. The workshop provided further insight into some of the best Japanese practices, and site visits to various monuments in Kyoto demonstrated how innovative fire protection systems are implemented.

Having extensive working experience with Cultural Heritage without Borders Albania (CHwB) at the World Heritage Site of Gjirokastra[1] proved to be the perfect foundation for developing the idea into an actual project. CHwB has been actively present in Gjirokastra since 2007, making significant contributions to the preservation of the city's cultural heritage. With financial support from the Prince Claus Fund and the Swedish International Development Agency (SIDA), a pilot project was implemented in one of Gjirokastra's first-category monuments.

A Brief on Gjirokastra – The Site

Gjirokastra, a small city in Albania, is a remarkable example of 18th- and 19th-century urban architecture. The Gjirokastra castle, which overlooks the entire valley, serves as a testament to the core of the settlement that expanded and developed into eight residential quarters and a bazaar.[2] Today, these areas form the historical center of the city. Gjirokastra was added to the UNESCO World Heritage list in 2005.

DOI: 10.4324/9781003356479-10

Spread over 6.85 hectares, the historical centre includes approximately 1200 vernacular houses within its borders, out of which 395 are designated as first and second-category monuments. Due to its cultural and historical wealth, Gjirokastra was declared a museum city in 1961 (Hadzic & Mamani, 2015).

Analysis of Threats and Vulnerabilities

The built heritage of Gjirokastra, located in a high-risk seismic zone, is vulnerable to a range of natural and human-induced disasters. Wooden ties traditionally reinforced the stone masonry of the buildings to resist possible earthquakes. However, there is no recorded evidence of damage due to seismic activities. The poor maintenance of the buildings or inappropriate restoration interventions throughout the years has negatively affected the condition of the traditional earthquake-resistant schemes. As a result, the buildings are more vulnerable to the damaging threat of earthquakes. Additionally, natural calamities like landslides are also threatening specific areas of the historical centre with an increased potential risk due to climate change, deforestation, and urban development. "Disaster Risk Management of Cultural Heritage Sites in Albania", Within the frame of the project: "Building Capacity in Natural Risk-Preparedness for Cultural Sites in Albania, United Nations Albania in collaboration with ICCROM and UNESCO, 2014.

Unresolved ownership issues and increasing emigration of residents are the primary reasons for many abandoned buildings across the historical centre. Abandonment results in a lack of maintenance, and additional factors like insufficient funds further lead to a rapid deterioration observed throughout the city. The lack of maintenance and inappropriate interventions also threaten the authenticity and integrity of the historical buildings (Mamani, 2014).

Among the many threats, fire is the one with a higher probability of occurrence and causing devastating effects. Throughout the city's history, entire houses or building complexes have succumbed to fire. The city centre, the bazaar of Gjirokastra, has been destroyed by fire twice, and even today, fires continue to destroy the monuments. The Koloi house, a first-category monument, caught fire in 2008, and with no reconstruction efforts, it is left today as a pile of stones. The Kokalari house, another monument, was burned down in 2014, only a few months after the finalisation of restoration work. Many other buildings suffered the same fate. The most recent one is the Topulli house, which burnt down in December 2019. The fire destroyed the house, but fortunately, the fire brigade prevented the fire from spreading to the adjacent house.

Several internal and external factors increase the vulnerability of the buildings towards fire. Old and poor electrical installations are often fire ignition sources. Due to the buildings' wooden combustible structures and interiors, a mere spark of fire can spread rapidly. Abandoned houses across the city, ungrounded electrical cables, overgrown vegetation, and scattered garbage are at significant risk of fire.

The firefighting infrastructure in the city is insufficient, with several factors limiting its effectiveness. Fire detection services are almost unknown, and the number of hydrants is insufficient to cover the neighbourhoods. Limited water supply, especially during dry summer months, leads to further insufficiency. The urban configuration of the city is also an obstacle in itself. The cobblestone streets of the historic quarters are often too narrow for fire trucks, while the close proximity of houses raises the risk of fire spreading rapidly from one building to the next. Furthermore, local residents lack awareness and training about fire and protection measures, such as prioritising saving lives over saving historic buildings.

Water Cisterns – A Pilot Solution Study

Historical cisterns are an integral element of Gjirokastra's historical dwellings. Due to the limited water resources, people have found an ingenious solution for collecting rainwater in dedicated rooms within these buildings. The water is collected through a system of gutters leading them to a large cistern where the water can be stored for a long duration. There are approximately 500 cisterns across all historical houses, with water capacity varying from 50 to 120 m^3 each (Mamani & Merxhani, 2016).

The city water supply system has affected the operational usage of these cisterns. Nowadays, most cisterns are unutilised and converted to storage and spare rooms. Since the city does not have sufficient water supply and climate change is accelerating its effects on freshwater resources, it is crucial to consider alternative usage solutions that do not necessarily require drinking water.

Cultural Heritage protection is important. Water cisterns represent a historical element at risk of extinction due to a change in the original usage of the space and the loss of the intangible heritage that accompanies its maintenance and functioning rituals. Water cisterns were a core element of a building's structure and people's life.

Monitoring water levels during monsoons, cleaning cistern interiors, and maintaining each element of the system, right from the gutters and pipes supplying the rainwater to the channels that pump water outside the house, were part of people's daily routine. The cistern built a social connection for members of large families and neighbours. Building a cistern in the house was a costly endeavour. Hence, affluent family homes with large cistern spaces allowed their neighbours to use the water. The Skenduli house represents an example of this practice. Two taps are installed in this house, one in the interior of the building for family use and another on the exterior wall for the neighbours. Interestingly, the exterior wall tap was built at a level higher than the interior one. This represents that the family was careful to secure a certain amount of water for their use in case of long drought periods. Growing families have led to separating big houses into two independent parts. However, the usage of water cisterns was always allowed for both families, even where the position of the cistern belonged to one half of the building.

Preserving tangible and intangible heritage in our historical towns is crucial. However, it is important to find ways of utilising them without compromising the values of the place. Natural resources also need to be used carefully and efficiently. Considering these very important aspects, finding alternative solutions that will help preserve the values of the space but also make them useful for the recent needs is very necessary and should be fully supported in the case of Gjirokastra historical town, there are numerous historical cisterns still filled with rainwater, but this water remains unused and hidden within the walls of the cisterns. The initial purpose of cisterns has diminished in recent times. However, cisterns can be a crucial element for protecting historical buildings. An alarming difference exists in the water capacity of a cistern and a fire truck. The capacity of a small cistern is 15 times larger than a single fire truck (Mamani, 2017).

Therefore, converting historical cisterns into water storage systems can potentially solve fire threats.

The project was initiated within the organisation CHwB Albania, which recognised the idea's potential to be in tandem with the organisation's objectives. A team was assigned, and the process began with a site survey and identifying possible cisterns for a pilot project. The selection criteria were based on the following listed aspects:

- Selecting cisterns in each of the neighbourhoods within the historical centre.
- Monument's historical value and importance.
- The presence of functional water cisterns in the monuments, nowadays, many cisterns remain unused or have a completely different function.
- The possible number of surrounding monuments that the fire extinguisher system could cover in case of fire.
- The accessibility of the monument by car, as most vulnerable monuments were those inaccessible by fire trucks.
- Water supply in the neighbourhood, since this is directly related to the usage of the hydrants, where these were available.

The survey identified 18 monuments with cisterns for possible inclusion in the project (Figure 8.1). A database was created for each of the selected monuments, including maps of their position, plans with the position of the cisterns, photos, detailed scenarios of the building access, and coverage of surrounding buildings, including historical monuments.

Several meetings were organised with the monument owners to gauge their willingness and ability to participate in the programme. The project proposal was also presented to the city's Fire and Rescue Department and the Regional Directorate of National Monuments, responsible for protecting monuments. Both institutions supported the idea and contributed with their expertise in the process.

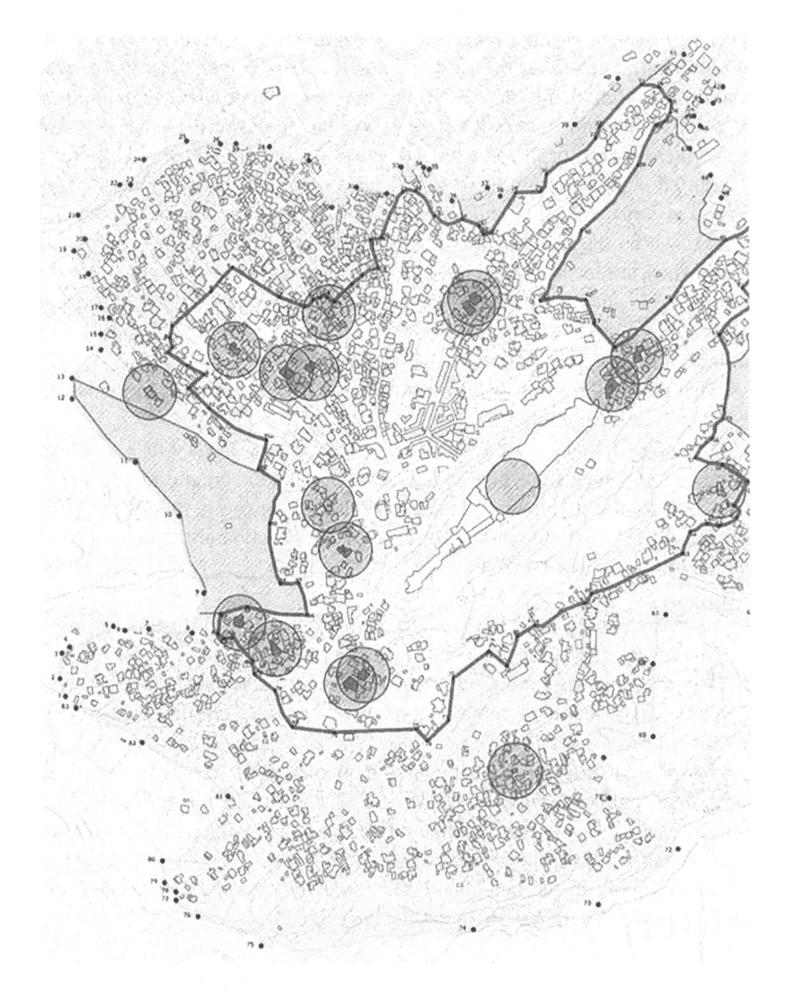

Figure. 8.1 Map of identified water cisterns within the historical centre.

Source: Author

The 18 selected monuments were analysed based on the building's condition and usage, the condition of the cistern and the gutter system, the volume of the cistern in the building, and ease of access to the hydrant.

ChwB Albania developed a project concept along with a mechanical engineer. The concept solution involves minimal intervention, retaining the historic fabric alongside the original function of the cistern.

Figure. 8.2 The hydrants.

Source: Author

Throughout the process, the CHwB team organised multiple meetings with different stakeholder groups, including monument owners, local institutions responsible for protecting monuments, the municipality, fire brigade, and specialised engineers (Figure 8.2). The team also received valuable inputs from field experts.

Based on the potential of each site, three different scales of proposals were developed:

- A firefighting system with single building coverage.
- A firefighting system installed in one building and also covering surrounding buildings in a radius of 30 m.
- A larger firefighting system where several cisterns could be connected to a network of hydrants enabling extensive neighbourhood coverage.

Consequent to discussions with the stakeholders and the availability of funds, it was decided to experiment with one pilot project. Under this project, one water cistern was converted into a fire suppression system covering the building where it was installed and the surrounding buildings in a radius of 30 m. The pilot project was implemented with the Prince Claus Fund and SIDA funding.

The aim of the project was twofold, revitalisation of the historical water cisterns as a fire prevention system and preservation of tangible and intangible heritage by preserving the historical fabric and the ritual of water collection.

Converting a Water Cistern Into a Fire Prevention System

Jaho-Babaramo[3] house, a first-category monument positioned in a densely built neighbourhood of the historical centre, was selected for the pilot project. The house is of a typology of two-winged building. The exact construction date is untraceable but based on the two-winged typology composition elements of the building, it is dated at the end of the 18th century.[4] The house is a three-floor structure where the cistern is positioned on the ground floor of the north wing. A network of gutters in the roof collects the water, and through a vertical pipe, it is directed to the cistern's interior. The cistern is a rectangular-shaped room built-in masonry structure covered with a stone vault. The walls in the interior are covered with traditional plaster, which has waterproofing properties securing the structure from humidity. Usually, the cisterns are placed in a deeper terrain for the ground, to keep the load and pressure that water transmits to the walls. If the terrain does not allow this, reinforcement buttresses can be added to the exterior of the walls.

Most cisterns have a manual system to control the water overflow that could threaten the stability of the structures. A small window in the sidewall of the cistern allows an entrance space for cleaning and maintenance purposes and acts as a control window for monitoring the water level. In some cisterns, there is a small hole under this window from where water can overflow, alerting the owner to disconnect the pipe that transports the water to the cistern. In the pilot project, this system is automatic, a rare example in the city. The maximum amount of water is controlled by a pipe placed inside the wall that discharges the water when it exceeds the maximum allowed level. This

Figure. 8.3 The final test.

Source: Author

pipe transports the water outside the building through underground channels (Mamani & Merxhani, 2016).

The cistern of the Jaho-Babaramo house has a maximum water capacity of 100 m^3. The intervention included small restoration work using traditional materials, techniques, and the design of the firefighting protection system. The final intervention included installing a submerged pump inside the cistern, which then was connected to a hydrant outside.

The protocols to be followed in case of fire were identified through discussions with the fire brigade specialists. The first action is to cut off the electricity in the area. Therefore, an automatic system was designed to function through an oil-powered generator, allowing it to work independently of electrical power.

While undertaking the course in Japan and discussing with the fire brigade team in Gjirokastra, it was concluded that an immediate reaction to fire is crucial. Hence, the system design included installing two hydrants in the exterior of the house: one with lower water pressure that can be used by the monument owner and the community living nearby, and the second that the fire brigade can use.

As a result, monument owners can promptly react by sustaining the fire spread and controlling it until the fire brigade arrives. Upon arrival, the firefighters can connect their pipes to the system and continue extinguishing the fire. The fire brigade can also use the cistern's water in case of fire in surrounding buildings.

The photos earlier show the final system testing carried out in collaboration with the fire brigade and the monument owners. A communal effort is an important aspect that must not be neglected while implementing the system. Since the people residing in these houses are the primary responders, creating awareness and training them for prompt action is crucial.

Therefore, during system testing, the firefighters also trained the community to use the hydrants in case of fire.

Conclusion

The pilot project implemented in Gjirokastra represents that traditional rainwater collection could be transformed into an effective fire prevention system. The number of cisterns is sufficient to cover an entire historical neighbourhood, but the focus must be on ensuring funding to support a large-scale intervention. The city's water supply system is improving. However, freshwater should be maintained and used only as a backup.

Many studies worldwide are considering the integration of cultural heritage in disaster risk reduction strategies and climate actions. The cistern project demonstrates that the perception of cultural heritage preservation can change from only preservation actions to more comprehensive sustainable solutions that can promote the creative usage of natural resources and utilisation of local knowledge in disaster risk reduction actions. Efficient use of natural resources is the need of the hour, and preservation of cultural heritage is vital for the local community's development. Therefore, such projects are vital to understanding and exploring innovative solutions that can contribute to the future sustainable preservation of cultural heritage and the planet.

Notes

1 Gjirokastra is a historic town located in the Drinos River valley in Southern Albania. It was originally constructed by prominent landowners around a 13th-century citadel. In 2005, Gjirokastra was designated as a UNESCO World Heritage site, along with Berat, another historic town in Albania, which was added to the same list in 2008. Both towns are considered rare examples of Ottoman-era architecture and character. Their proximity to each other and location in the Balkans demonstrate the rich and diverse urban and architectural heritage of the region. They bear witness to a way of life that has been influenced over a long period by the traditions of Islam during the Ottoman period, while at the same time incorporating more ancient influences.

2 Bazaar was the marketplace in the traditional towns. In the case of Gjirokastra, it is a whole quarter positioned in the crossroads that connect all historical neighbourhoods of the city and contains rows of little shops where people can buy food, clothes, and shoes and can find all kinds of artisans. Part of the bazaar is also a Mosque, and many of the shops belonged to the Muslim community who are financially supporting its maintenance.

3 The buildings are named as per the surname of the family who owns them.

4 Riza, E. (1981). *Qyteti – Muze i Gjirokastrës (monografi)*. Tiranë: Shtëpia Botuese 8 Nëntori.

References

Hadzic, L., and Mamani, E. (2015). Condition Assessment for Monument Buildings in Gjirokastra. Paper Presentation. *Monuments, Proceedings of 50 Years Study and Restoration of Albania Cultural Heritage*, 54(2). Institute of Monuments of Culture in Albania.

Jigyasu, R., and Arora, V. (2014). *Disaster Risk Management of Cultural Heritage in Urban Areas – A Training Guide*. Kyoto: Research Center for Disaster Mitigation of Urban Cultural Heritage, Ritsumeikan University (RitsDMUCH).

Mamani, E. (2014, September 6–22). *DRM Plan for Historic Center of Gjirokastra*. Paper Presentation. International Training Course on Disaster Risk Management of Cultural Heritage, Ritsumeikan University, Kyoto Japan.

Mamani, E. (2017, August 28–September 16). *CHwB Albania Projects on Risk Management and First Aid Response*. Paper Presentation. International Training Course on Disaster Risk Management of Cultural Heritage, Ritsumeikan University. Final Report on B+CARE Program, CHwB Albania, Kyoto, Japan.

Mamani, E., and Merxhani, K. (2016). *Water Cisterns in Historical Houses*. Paper Presentation. Monuments, Proceedings of ISCCE Conference 2015, 53. Institute of Monuments of Culture, Albania.

Riza, E. (1981). *Qyteti – Muze i Gjirokastrës (monografi)*. Tiranë: Shtëpia Botuese 8 Nëntori.

9 Participatory methods for DRR in Santo Domingo Tehuantepec, Mexico

David A. Torres Castro

Introduction

Santo Domingo Tehuantepec (Tehuantepec) is a 500-year-old historical city in the south-east of Oaxaca, Mexico, with a population of around 64,000. It is a regional, cultural, and political epicentre today. The city has an irregular urban layout divided by the Tehuantepec River, developed through a century-old transformation process. There is urban development around the main square, where many historic buildings are still preserved. Most of them are built on masonry, adobe, and timber, in continuity of the traditional construction techniques (Torres Castro, 2019). The urban structure clusters 14 neighbourhoods today, locally known as *barrios*, each with its cultural personality, social organisation, and function within the city's urban system. At the heart of every *barrio* lies a historic chapel dating back to the 17th century, serving as a local cultural, religious, and civic focal point. Many *barrios* use the chapel's atrium as a multipurpose public square for street markets and public assemblies.

Tehuantepec is located in one of Mexico's highest seismically active regions, with a history of strong earthquakes that damaged the local heritage and the built environment (Torres Castro, 2018). According to Mexico's National Seismic Service database, over 100 earthquakes of magnitude five or above have been registered in the last decade (SSN, 2019). Additionally, the city's location near the southern Mexican coast adds other hazards, such as tsunamis and tropical storms. In 2018, Mexico's Meteorological Service registered 25 cyclones on the Pacific coast, 12 were classified as hurricanes, and 9 were of category III or above (Bravo Lujano, 2019). In the last few years, these events have doubled above the annual average, representing the gradually increasing frequency of cyclones and heavy rainfall, most likely due to climate change (Kelman, 2020).

DOI: 10.4324/9781003356479-11

Main Objectives and Goals

The pilot project originally intended to improve the protection of heritage assets stored in the 16th-century building and former Dominican Convent in Tehuantepec. In 2017, after the earthquakes, a participatory Disaster Risk Management (DRM) plan was designed for the site and proposed as the primary goal. The project was founded on the principle that both the Convent and the collections guarded inside are active agents of identity and social cohesion. In an emergency situation, these kinds of assets can facilitate psychosocial strengthening of local society as long as they remain operational during both the response and recovery phases (García Souza, 2019). Hence, the project proposed a workflow based on participatory activities, such as risk mapping and vulnerability assessments designed in collaboration with local stakeholders. The outcome would be to raise risk awareness among the local community, identify risk hot spots, strengthen response capacity, and reduce the impact of potential risks that may threaten the city's heritage assets.

Materials and Methods

The first phase of the pilot project was based on the combination of qualitative and quantitative data. It included the gathering of local information regarding threats and vulnerabilities through participatory methods such as risk mapping and surveys and was complemented with quantitative data from official sources such as the National Centre for Disaster Prevention's (CENAPRED) Risk Atlas. The combination of both types of data allowed for triangulation and validation of information.

(1) Stakeholder Analysis

Working collaboratively with key stakeholders was essential during the workshop and the logistics and processing of the data, as they have the knowledge to perform an assertive and realistic vulnerability and risk assessment. Therefore, the team worked closely with the project's local partner as an entry point identifying other relevant participants and stakeholders for the participatory activities. Thus, a stakeholder map was designed, including institutional, private, and community stakeholders.

The institutional stakeholders included state representatives from the National Institute of Anthropology and History (INAH) due to their authority regarding cultural heritage. Additionally, the head of the built heritage office from the local government was included, along with the volunteers from the civil protection body. The private sector included a representative from the company responsible for the Convent recovery. Furthermore, community members, mostly elders due to time availability, and the local chronicler were invited. The project's local partner and an experienced architect in the local heritage conservation sector were also invited.

(2) Methodology and Scope

The first approach entailed working with local authorities and key stakeholders to create participatory maps of attributes and threats to lay the foundations and tools for a DRM plan specifically for the 16th-century building and the collections it stores. However, during the preparatory desk work, a strong relationship between the Convent and the surrounding buildings was identified, becoming counterproductive to work with them individually. As a result, the project was expanded from its original scope to cover the first perimeter within the historical centre of the city. The new area covered most historical buildings, working as a cluster of continuous built spaces, including housing, streets, and open spaces. This proved beneficial as many potential risks and solutions were identified in the surrounding buildings and spaces, not from the Convent building.

(3) Preparatory Work

Before the fieldwork, maps were retrieved from Google Earth (2022), to prepare a detailed map of the city's street layout as a base layer for information entry. Additionally, satellite images and GIS files were retrieved from CENAPRED's Risk Atlas (2018) and used to generate a basic risk profile for the study area. Furthermore, logistic preparations were done in collaboration with the local partners to find a suitable space for the workshop.

(4) Survey on Cultural Significance

A short survey was prepared, piloted, and applied between July 24 and 25, 2018, focusing on the cultural significance and risk perception of local cultural heritage. The project team surveyed a random population sample in Tehuantepec's main square.

(5) Participatory Risk Mapping Workshop

In September 2018, a three-day workshop on 'Participatory Mapping of Risks for Cultural Heritage Towards the Construction of Resilience and Co-responsibility' was conducted in Tehuantepec. There were 16 participants, including local stakeholders and authorities, local civil protection staff, and private company actors involved in the recovery process after the 2017 earthquakes. The workshop focused on recognising the attributes of cultural assets, identifying the risks and vulnerabilities related to those assets and problem-solving mapping through collaborative tools such as the Disaster Imagination Game methodology (Delica-Willison and Gaillard, 2012) (see Figure 9.1). Participants were divided into three groups, with representatives from all sectors; all were assigned with equal tasks to compare and find differences and similarities between them. The team members acted as facilitators. After the mapping activities, a risk and vulnerability matrix was built in collaboration with the participants, containing information from the maps and discussions.

After the workshop, the information gathered was triangulated by the team members through urban context documentation and assessment, field observation, and validation of identified risks (Ahmed et al., 2019).

(6) Data Processing

Following the workshop, the team members digitised the obtained maps, integrating vulnerabilities, risks, cultural assets, and mitigation activities from the three teams into one map. The processing was done in collaboration with Mexico's National School of Conservation, Restoration and Museography (ENCRyM).

Figure 9.1 Disaster Imagination Game tool used to identify threats, vulnerabilities, strengths, and mitigation strategies during the participatory workshop with local stakeholders.

Source: Author

Results and Discussion

(1) Cultural Heritage Attributes and Vulnerability Mapping

The project team identified that poor cultural policies and practices at the local level, along with poor urban and waste management, have resulted in a lack of maintenance of heritage assets and subsequent deterioration of historic structures. This led to increased structural vulnerability within buildings in the selected perimeter. Other hazards such as water ingress due to floods and rainwater filtration among buildings also contributed to the dire state of conservation of historical structures. These risks were poorly understood and mitigated and added to the region's high seismic activity that continuously threatens cultural assets. Altogether, structural instability associated with lack of maintenance or occupancy for some buildings represents a risk of collapse. Furthermore, fire due to electrical failure, combined with poor waste disposal, was also identified as a major risk surrounding cultural assets (see Figure 9.2). In addition to building collapse, flood risk, however low, was identified and mapped as a potential risk that may affect emergency response procedures by blocking access to affected areas, particularly near slopes and riverbanks. In this regard, main access and exit roads were identified and marked as a priority in the overall emergency management procedures of the city, including removing debris among essential activities.

It is worth mentioning that structural vulnerabilities from a group of already restored historical properties used for residential purposes were also listed by the participants, remarking the risk of partial or total collapse. This is important as it reflects the perception of risk that could lead to mistrust regarding traditional construction systems; in this case, it shows local members are sceptical of the resistance of adobe buildings (*cf.* Cortés Robledo and Cruz Gutiérrez, 2014). While this may change as the restoration of damaged homes progresses and dwellers experience living in a recovered property, it is relevant as it mirrors poor communication policies and plans from the national or local cultural authorities.

(2) Intangible Cultural Heritage

Mapping routes of cultural importance is essential since they are recognised as a fundamental part of religious celebrations and are therefore agents for social cohesion. These represent an important element to consider in the DRM plan, especially in the reconstruction and recovery stages, where modifications or alterations of the urban layout may interfere with them. While excluded in this pilot project, efforts towards protecting intangible heritage routes will have to be included in the following stages of the project, avoiding its modification at all costs as they are a fundamental part of the cultural activities of the city's historic centre.

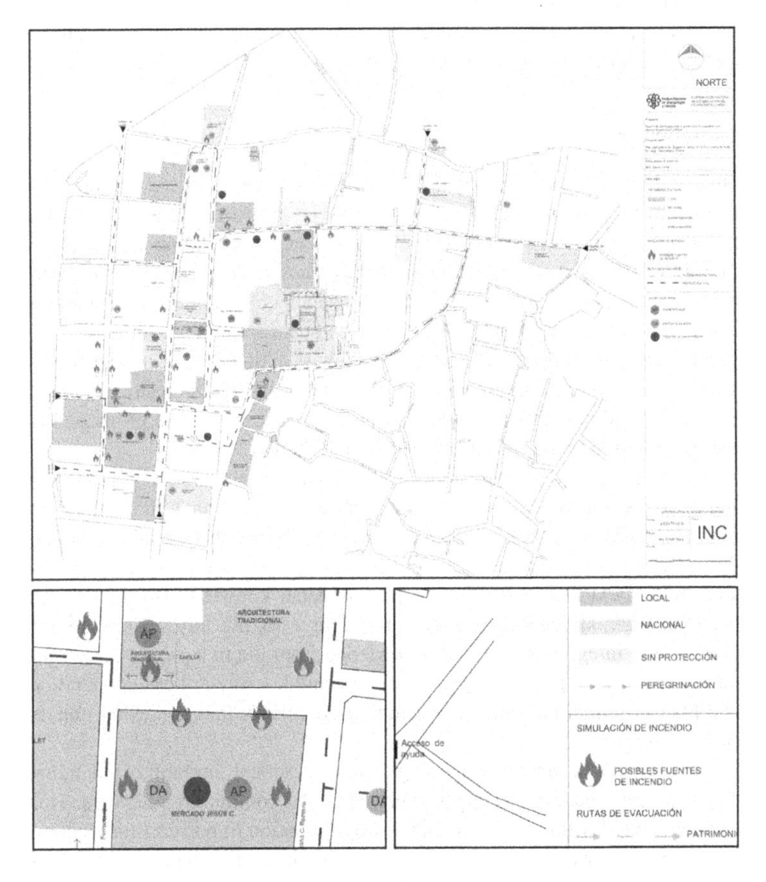

Figure 9.2 Vulnerability and risk maps generated through participatory tools. Processed and digitised by Citlalli Silva with data from the workshop.

Source: Author

(3) Capacity Building and Training

Lastly, capacity-building tools were recognised as essential strategies to provide adequate assistance and protection to heritage assets in emergency situations. These are matched with the need to create a local first-response task force focused on cultural heritage.

In response, collaboration with local authorities and the community was proposed to designate public squares as evacuation areas for affected heritage assets in emergencies. This initiative includes training a specialised team and preparing a toolbox with essential equipment for these tasks that should be kept in a space with quick access. Additionally, it was observed that the absence of a firefighting station in the city weakens the local response capacity.

Conclusion

During the participatory mapping workshop, two areas of great cultural relevance were identified as high-risk spaces to be prioritised: the Main Market and *Benito Juarez* Elementary School, and the area surrounding the Cathedral of Santo Domingo atrium, including the *Mercado Campesino* (farmers' market). Together, this cluster forms half of Tehuantepec's main historic buildings and represents the city's most culturally significant areas, as confirmed by the survey. The risk of fire was identified as highest in these areas, mainly due to a large concentration of informal trading posts, improvised electrical wiring, and significant amounts of rubble and garbage. These must be prioritised in the forthcoming phases of the construction of the DRM plan.

Direct mitigation measures must be undertaken collaboratively by various stakeholders, including community leaders and institutional participants, to reduce risks over cultural heritage. Hence, considering the use of a combination of qualitative and quantitative techniques, such as the risk map generated by CENAPRED with the mapping of local knowledge of risks, is an essential method for understanding the degree of the city's vulnerability. Community integration and local knowledge from the first phases of designing an emergency plan are undoubtedly essential. It strengthens cooperation and fosters mutual responsibility among all involved, thus achieving greater sustainability.

Notably, the selected methodologies have practical limitations, including the representativeness and engagement of all the community members. This became evident while analysing the low participation of non-institutionalised members in the workshop. Moreover, it is essential to highlight the difficulties of integrating the results of qualitative information into the actual recovery activities undertaken after the 2017 earthquakes. This may be related to the prevalence of techno-centric, object-based approaches within the institutional and cultural systems. Lastly, proceeding with an inclusive and culturally diverse disaster risk management policy is recommended.

Acknowledgements

The author would like to thank Paola García, Luis Mario Díaz, and Mitzy Quinto for contributing to the project, and Citlalli Silva for the digitalisation and processing of the maps.

References

Ahmed, B., Sammonds, P., Saville, N.M., Le Masson, V., Suri, K., Bhat, G.M., and Thusu, B. (2019, April 2018). Indigenous Mountain People's Risk Perception to Environmental Hazards in Border Conflict Areas. *International Journal of Disaster Risk Reduction*, 35, 101063. https://doi.org/10.1016/j.ijdrr.2019.01.002.

Bravo Lujano, C. (2019). *Resumen de la temporada de ciclones tropicales del año 2018*. México: Comisión Nacional del Agua (CONAGUA).

Centro Nacional de Prevención de Desastres (CENAPRED). (n.d.). *Atlas Nacional de Riesgo*. Sistema Nacional de Información Sobre Riesgos. www.atlasnacionalderiesgos.gob.mx/archivo/visor-capas.html.

Cortés Robledo, B., and Cruz Gutiérrez, F. (2014). El Chalet de Juana "Cata" Romero: Un monumento artístico reminiscente en el Istmo de Tehuantepec, Oaxaca. *La Revista Del Instituto Del Patrimonio Cultural Del Estado de Oaxaca*, 27. http://todopatrimonio.com/pdf/GacetaINPAC/Gaceta27.pdf.

Delica-Willison, Z., and Gaillard, J.C. (2012). Community Action and Disaster. In: *Handbook of Hazards and Disaster Risk Reduction*. London: Routledge, pp. 711–722. https://doi.org/10.4324/9780203844236.ch59.

García Souza, P.P. (2019). *Taller sobre patrimonio en San Mateo del Mar, Oaxaca*. Working Document. Mexico: INAH.

Google Earth Pro 7.3.3.7699 (64-bit). (August, 2022). Map of Tehuantepec urban area. Lat. 16*20'1.92"N, Lon. 95*13'50.52"O, Eye alt 906m. Airbus 2023. <https://earth.google.com/web/search/Tehuantepec,+Oax./@16.3333979,-95.23047355,55.94021609a,871.5212171d,35y,-0h,0t,0r/data=CnwaUhJMCiUweDg1YmZmMDYzODcxYmI2MDc6MHg2MmYxY2YyZjczYWEzZjllGVZwiW2cUjBAIb01sFWCz1fAKhFUZWh1YW50ZXBlYywgT2F4LhgBIAEiJgokCRBXCqjDZzRAEQtXCqjDZzTAGYFIYUtRPEFAIX_ZDmBy4lDA> (Accessed September 10th, 2022).

Kelman, I. (2020). *Disaster by Choice*. Oxford: Oxford University Press.

Servicio Sismológico Nacional. (2019). *Catálogo de sismos*. http://www2.ssn.unam.mx:8080/catalogo/.

Torres Castro, D. (2018). The Dominican Convent of Tehuantepec, Mexico. A Disaster Risk Management Draft Plan. In: Jigyasu, R., and Kim, D. (Eds.), *Proceedings of UNESCO Chair Programme on Cultural Heritage and Risk Management International Training Course on Disaster Risk Management of Cultural Heritage*. www.r-dmuch.jp/en/results/dl_files/Proceedings_of_ITC_2013.pdf.

Torres Castro, D. (2019). *El Convento Dominico de Santo Domingo Tehuantepec, Oaxaca. Proyecto piloto para el diseño de un plan participativo de Gestión de Riesgos para Patrimonio Cultural*. Working Document. Mexico: INAH.

Part 3

DRM Plan Implementation – Stakeholders' Participation/ Decision-Making

10 Risk assessment of Humberstone & Santa Laura Saltpetre Works

Marcela Hurtado

Introduction

The Disaster Risk Management Plan (DRMP) for Humberstone and Santa Laura Saltpetre Works was developed in the framework of an agreement between the Technical University Federico Santa María (USM) and the National Centre of World Heritages Sites (CNSPM), which reports to the Chilean Ministry of Culture, Arts, and Heritage. Funding was obtained from the World Heritage Centre to develop the project on Strengthening Disaster Risk Management at three World Heritage Sites in Chile. The project was executed between 2017 and 2019.

In 2005, the project site Humberstone and Santa Laura Saltpetre Works was registered on the UNESCO World Heritage List and added to the World Heritage List in Danger due to its fragile conservation state. The site illustrates an important part of Chilean and world economic history due to its location in the Atacama Desert. One of the largest nitrate deposits in the world are found here. From 1850 until 1930, more than 200 company towns, called Saltpetre Works, were established in this area. Extreme landscape and distinct social interaction integrated by North American and British mining entrepreneurs, Chilean workers from different regions of the country, and Peruvian and Bolivian workers originated a local identity – *pampina* – that still prevails. However, most of these settlements disappeared with the definitive closure of industrial activity in the mid-1980s. Others have been abandoned or dismantled by their last owners (González Miranda, 2002; Garcés Feliú, 1999; Macuer Llaña, 1930).

The site comprises two complementary components – Humberstone and Santa Laura – each former Saltpetre Works. Humberstone contains the best-preserved remains of the facilities for workers and owners, represented by different housing typologies and public equipment such as theatre, school, church, market, and hospital, among others. Daily life in the Saltpetre Works can be understood from these facilities, that is strongly marked by social segregation and the sacrifice of workers and their families. On the other hand, Santa Laura has the best-preserved ensemble of industrial

DOI: 10.4324/9781003356479-13

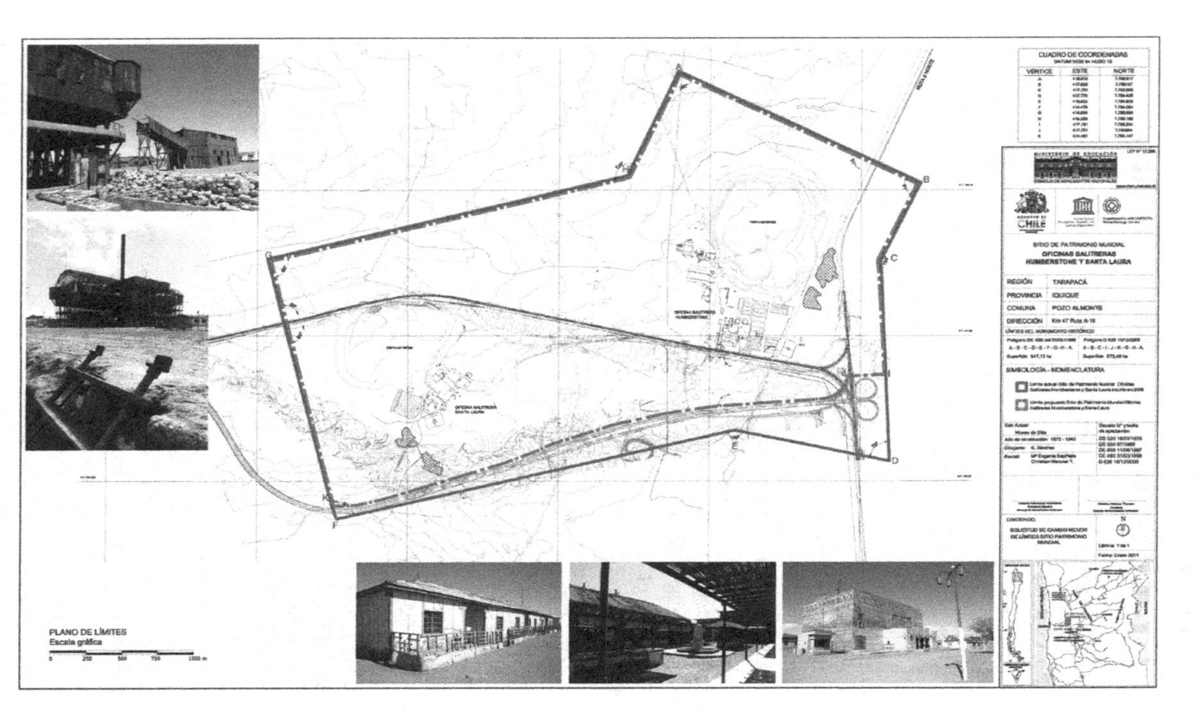

Figure 10.1 World Heritage Site Humberstone and Santa Laura Saltpetre Works, Pozo Almonte, Chile.

Source: Map, whc.unesco.org; Photos by the author

facilities, including a unique leaching plant and other buildings and industrial equipment that provide knowledge of the production process, from the extraction of the mineral to the production of Saltpetre (see Figure 10.1).

The initiative to develop a DRMP for the site started from the research developed by the USM in risk assessment in cultural heritage sites and the interest of the CNSPM to develop tools to reinforce the conservation and maintenance of Chilean World Heritage Sites. At the same time, the State Party was updating the management plan for the site and some conservation projects as part of efforts to remove the site from the World Heritage List in Danger. Notably, after 14 years, in 2019, the site was removed from that category.

The private Saltpetre Museum Corporation (CMS) manages the site. CMS was incorporated in 1997 by the former inhabitants of the Saltpetre Works to preserve the memory of these places. This group played an essential role in promoting the nomination of the site for world heritage. Currently, CMS has 296 members, 256 founders, and 27 staff members developing various tasks focused on the preservation and advertising of the site.

Risk Management Plan

CNSPM coordinated and conducted their process (see Figure 10.2). CNSPM was a crucial intermediary between the site manager and other key stakeholders. They also organised two workshops to disseminate project results in different phases. Three phases were defined to develop the DRMP considering

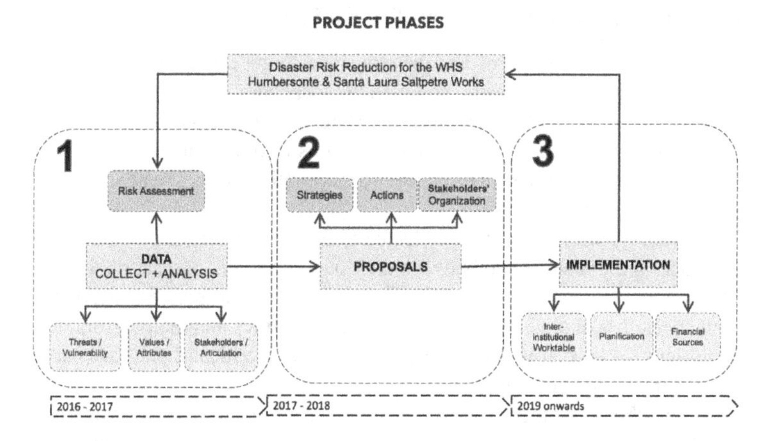

Figure 10.2 Methodology for the development of the disaster risk management plan.

Source: Author

the recommendations and guidelines from international organisations dedicated to cultural heritage preservation (UNESCO et al., 2010, 2012; ICCROM et al., 2016) and some specialised literature review (Jigyasu, 2005; Ravankhah et al., 2014, 2017; Davies et al., 2015).

The first phase focused on data collection and analysis to obtain basic information from the site like values and attributes, main hazards, and stakeholders. The information enabled defining a matrix for risk assessment for each hazard. Furthermore, it helped obtain a preliminary outline of the relationship between different stakeholders and their specific responsibilities associated with disaster risk management. In addition to the historical and documentary review for this phase, a workshop was held with the participation of *pampinos*, the site manager, experts, and other key stakeholders.

The second phase consisted of preparing proposals for disaster risk reduction derived from the results obtained in the first phase. The proposals were grouped into two categories: strategies and projects. Strategies refer to programmes and other initiatives to be organised in coordination with public institutions to improve disaster risk management and contribute to the plan's implementation. On the other hand, projects are specific interventions on the buildings, infrastructure, or public space in the site, for risk reduction and protection of staff and visitors.

The third phase corresponds to the presentation of the results and the implementation. A workshop and coordination meeting with key local stakeholders was held on-field to establish a regional work group focused on the implementation of the plan. Due to the COVID-19 pandemic, the DRPM is currently stalled. It is expected to resume its activities during the first semester of 2022.

Critical Threats and Vulnerability Factors

The values and attributes of Humberstone and Santa Laura Saltpetre Works were validated through a workshop with the local community. Additionally, a nomination dossier and the retrospective Statements of Outstanding Universal Value for the site were also prepared. From this information, some areas and buildings of both components were prioritised for the risk assessment, in addition to their general analysis.

Moreover, information was collected to estimate the most critical threats considering the relationship between their occurrence and impact on the site. The same was evaluated through a review of historical records, interviews with specialists, and field visits to inspect the facilities and their surrounding conditions. Finally, it was concluded that the most critical threats are earthquake, fire, wind, and humidity.

For earthquakes, there is a history of occurrences affecting the buildings, which can be verified in both components. As in the rest of the country, the recurrence of earthquakes is high. In the case of fire, there is no precedent, but it was estimated that, due to the potential impact, the severity of the threat is very high. In the case of environmental pressures determined from the site's location, the recurrence is very high (daily), and the impact on the goods is low.

The risk assessment was executed for each threat. The vulnerability factors and their corresponding weightage were established by threat. Vulnerability factors were divided into four categories: intrinsic, context, management, and use. For each of these categories, secondary factors were defined. The latter were outlined based on the characteristics of the site, its buildings, public spaces, and the context. Extensive literature was also reviewed (Ferreira et al., 2016; Yen et al., 2015; Huang et al., 2009; Masi et al., 2014; Jiménez et al., 2018; Sesana et al., 2020; Forino et al., 2016). Intrinsic factors included data on materiality, height, and conservation state of structural and non-structural elements. The factors associated with the context included information on exposure or orientation, type of grouping, distance to access, distance to a pedestrian evacuation route, and vehicular accessibility. Management factors encompass questions about the disaster risk management plan, maintenance plan, fire detection systems, implemented security plan, and staff trained in disaster risk management. Finally, the factor associated with use comprises type, frequency, and the number of visitors or users.

In addition to the previous considerations regarding weighing each factor by threat, on-field work was undertaken to adjust the values. For defining the vulnerability indicators, future assessments to be done by the CMS staff and the site manager were considered. Hence, the method should fulfil the objective of providing credible risk index results without complexity. The indicators and weightage were then systematised in a matrix to facilitate data entry and obtain results for future risk assessments.

As part of the data collection, stakeholders, their implications with disaster risk management, and relations between them were identified. Stakeholders were divided into three groups according to the territorial scope of their functions, local or regional, national, and international.

Risk Assessment

During the first phase, a risk assessment was conducted for both components, focusing on the buildings and facilities defined as priorities according to their values. The risk index by threat was established, showing the incidence of each of the indicators. Even though some results were predictable (e.g. 'very high' risk index against fire in wooden buildings), it was important to have disaggregated data, that is, separated by vulnerability factors. Thus, it was possible to observe the high incidence of factors associated with management, mainly related to the existence of risk management plans, implementation of drills, or training of the field staff. As a result, it is possible to simulate scenarios with an implemented disaster risk management plan using the evaluation matrix. It is also possible to verify the rate of total risk index decrease for a specific attribute associated with a specific threat.

Theatre

Vulnerability factors	Earthquake	Fire	Wind	Humidity
Intrinsic	1,20	1,40	0,80	0,80
Context	0,10	0,50	0,60	0,60
Management	0,50	1,20	0,60	0,50
Use	0,05	0,30	-	0,10
TOTAL	**1,85**	**3,40**	**2,0**	**2,0**

Market

Vulnerability factors	Earthquake	Fire	Wind	Humidity
Intrinsic	1,80	1,40	0,60	1,00
Context	0,35	0,88	0,45	0,90
Management	1,00	1,20	1,20	1,00
Use	0,10	0,35	-	0,12
TOTAL	**3,20**	**3,83**	**2,25**	**3,03**

Workers' housing – Typology D

Vulnerability factors	Earthquake	Fire	Wind	Humidity
Intrinsic	2,10	0,70	0,80	1,60
Context	0,20	0,38	0,90	1,20
Management	1,00	1,20	1,20	1,00
Use	0,13	0,15	-	0,08
TOTAL	**3,43**	**2,43**	**2,90**	**3,03**

Casa de fuerza 2

Vulnerability factors	Earthquake	Fire	Wind	Humidity
Intrinsic	1,50	0,53	1,20	1,00
Context	0,15	0,38	1,05	0,60
Management	1,00	1,20	1,20	1,00
Use	0,13	0,15	-	0,05
TOTAL	**2,78**	**2,23**	**3,45**	**2,65**

Planta lixiviación

Vulnerability factors	Earthquake	Fire	Wind	Humidity
Intrinsic	1,50	1,40	1,40	0,80
Context	0,20	0,75	1,20	0,45
Management	1,00	1,20	1,20	1,00
Use	0,10	0,10	-	0,05
TOTAL	**2,80**	**3,45**	**3,80**	**2,30**

Figure 10.3 Risk assessment for some of the buildings in both components of the site.

Source: Author

In Humberstone, the results show high vulnerability, mainly to earthquakes and humidity. This is derived from the material and structural characteristics of the buildings, the state of conservation, the grouping system, and the lack of a comprehensive disaster risk management plan, including all threats. In the case of seismic risk assessment, the results of 'high' and 'very high' are associated with three factors: the state of conservation, the accumulated damage from previous earthquakes, and the construction system. The latter, known as 'pampino concrete,' has shown low load resistant capacity. The results showed two probable sources of humidity, the daily environmental humidity produced by the proximity to the sea and the geographical configuration. Interestingly, cold currents originating in the sea leave a cloud over the site nearly every morning for a few hours and therefore the materiality of the buildings is a key factor. On the other hand, humidity observed in some areas could be related to the use of sanitary facilities. Even though it is not an incident, it needs detailed verification. In any case, the problem derived from humidity manifests itself mainly in erosion at the base of the walls, which weakens the structure. Thus, the buildings are highly vulnerable to potential earthquakes.

Although there are few buildings with a 'very high' risk index against fire, it is important to note that these are emblematic and unique like the theatre, church, market, and school. These buildings correspond to the civic centre, one of the characteristic spaces of the company towns and one of the attributes of the site.

In Santa Laura, for the leaching plant, a one of its kind, the risk value for both fire and wind threats is 'very high'. The leaching plant is a monumental wooden structure with iron elements covered with metal plates. In Santa Laura, the risk from wind is generally high as most of these structures are clad in metal sheets and exposed to daily wind, especially in the afternoon. Hence, this primarily represents a risk to the safety of visitors and workers.

At the end of the evaluation phase, a workshop was organised to create awareness about disaster risk management at World Heritage Sites, the methodology, and the progress of the Humberstone and Santa Laura Saltpetre Works project. Cultural heritage professionals from public and private organisations and professionals from disaster risk management participated in this workshop, along with local decision-makers, such as councillors and public office staff members. The experience was highly positive. From an interaction session, an activity was developed where the attendees were divided into groups, where they defined threats and potential vulnerabilities for the World Heritage Sites. Finally, the groups presented the results, showing significant commitment and interest in this issue.

As previously mentioned, the second phase focuses on defining strategies and actions for disaster risk reduction based on the results obtained in the risk assessment and through the attributions and characteristics of the stakeholders. As per their scope, the strategies are classified into five areas: institutional, management, education and dissemination, financial resources, and infrastructure. Activities were defined for each area, along with the stakeholders in charge and their priorities. One of the activities defined as priority was to establish and train a regional inter-institutional working group for disaster risk management and heritage. For education and dissemination activities like distribution and promotion of the disaster management plan for the site within the community, training in disaster risk reduction for tour operators, and dissemination of safety protocols among visitors, especially schoolchildren, were prioritised. For infrastructure, the priority activities were implementing passive and active measures against fire, installing safety signs, and sanitation of the site, among others.

Areas	Activities	Stakeholder in charge	Priority (1 to 3)
Institutional	Integrate disaster risk management into the legal or regulatory instruments of each institution	ONEMI – SNGP	1
	Strengthen capacities for DRR in heritage in key institutions	ONEMI – GORE	2
	Strengthen inter-institutional networks for DRR in heritage	ONEMI – GORE – SNGP	1
Management	Lead DRRP implementation	CMS – SNGP – ONEMI	1
	Strengthen the capabilities of the Site manager team	CMS – SNGP – ICCROM	2
	Lead the articulation of the different actors in DRR of the Site	CMS	1
	Implement corresponding protocols	CMS + Local emergency services	2
Education/ Dissemination	Spread DRRP in the local community	Local media	1
	Training of key actors for the DRM of the Site (e.g. firefighters)	SNGP + Academia + International Cooperation	2
	Training of students in DRR	ACADEMIA	3
	Include DRR topics in WHS in schools, universities, and other institutions.	SEREMI Education	3
	Train tour operators in DRR for WHS	SERNATUR	2
	Training community	Emergency forces + Academia	3
Financial sources	Allocation of funds for the implementation of the DRRP	GORE	1
	Allocation of emergency funds	GORE – CMS	2
Infrastructure	Execution of emergency works	CMS – DA MOP	1
	Installation of safety signs	CMS	2
	Complete the cleaning of the Site	CMS	2

Figure 10.4 Proposal of activities for disaster risk reduction for Humberstone and Santa Laura Saltpetre Works.

Source: Author

The vulnerability assessment by building enabled defining a series of recommendations, systematised into specific actions, by building and by threat. The current and potential impact and threatened cultural values were indicated for each case. Intervention priority was also established. In some cases, it was recommended to repair a part of the structure or restrict visits temporarily. In other cases, monitoring environmental conditions such as wind or humidity to develop specific intervention projects was recommended. At this point, some coincidences were identified with the ongoing and future conservation projects. Although the development of this risk management plan was initially not executed in coordination with the site's conservation plan, attempts were made to integrate the measures that both proposed.

This phase defines the relationship scheme for disaster risk management among the different stakeholders (see Figure 10.3). In the proposal, few current practices were formalised with the site manager as the central figure. The National Centre of World Heritage Sites (CNSPM) was designated to establish relationships with other national and international organisations like the Ministry of Public Works, ICOMOS, and World Heritage Centre. The *pampinos would continue to* establish contact with other community groups and local media. The working group emerged as a new entity to unite stakeholders with competence in disaster risk management and cultural heritage from different institutions to build relationship with local institutions. While this group was not predefined at the beginning of the project, it emerged as an idea during various project meetings and activities with different stakeholders. Hence, the commitment and interest shown by these stakeholders, especially the regional authorities for culture and emergencies, were fundamental. Other

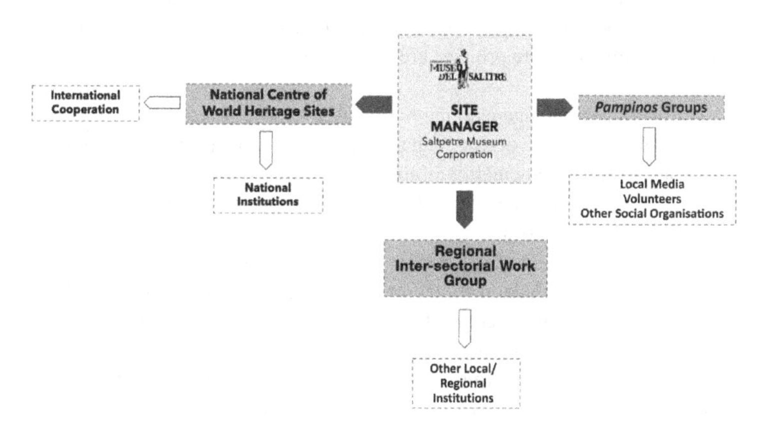

Figure 10.5 Articulation proposal between different stakeholders for disaster risk reduction in the site.

Source: Author

key stakeholders, such as firefighters, were also enthusiastic and collaborative with the idea.

The last phase corresponds to the implementation of the plan. Two unforeseen situations in the country made the implementation progress challenging. At the end of 2019, a series of demonstrations were triggered by social demands in the country, which altered the normal functioning of public institutions. On the other hand, in March 2020, restrictions on mobility and other activities were imposed due to the COVID-19 pandemic. However, the regional intersectorial worktable for disaster risk management in cultural heritage has been installed. This was led by the Regional Secretary of the Ministry of Culture and the Regional Director of the National Office for Emergencies. Other key local authorities from the regional government, municipality, firefighters, police, Regional Secretary of Tourism, and Ministry of Education also collaborated. The site manager is also a key participant in the first task of implementing the DRMP for Saltpetre Works. It is proposed to begin implementing the plan and use this experience to advance in consolidating the group's work and develop activities for disaster risk management in the remaining cultural heritage sites of the region during the first semester of 2022.

Conclusion

Developing the DRMP was an enriching and challenging experience due to a lack of experience in risk management planning, and limited knowledge on the matter applied to architectural and urban sites. The collaboration with the CNSPM, the site manager, and the local community was essential as they provided their experienced expertise and knowledge of the site and its issues.

Finally, some important aspects to highlight from this experience which will be helpful for future projects are as follows:

- The adaptation of a general methodology to a specific case.
- The importance of working with all stakeholders to have a strong commitment for the implementation of the plan.
- The role of the site manager as leader of the process, the need to empower, and install capabilities.
- The risk index results disaggregated by vulnerability factor will contribute to propose risk reduction strategies and measures among decision-makers.
- The importance to integrate the DRMP with other ongoing initiatives, especially maintenance and conservation plans.
- The dissemination of project results, along with the process and methodologies in forums, provides an opportunity for the different stakeholders to interact.

Acknowledgement

I would like to express my gratitude to all professionals from National Centre of World Heritage Sites, the team from Saltpetre Museum Corporation, and the students from the Department of Architecture at Technical University Federico Santa María for the collaboration and support during the project.

References

Davies, T., Beaven, S., Conradson, D., Densmore, A., Gaillard, J.C., Johnston, D., Milledge, D., Oven, K., Petley, D., Rigg, J., Robinson, T., Rosser, N., and Wilson, T. (2015). Towards Disaster Resilience: A Scenario-Based Approach to Co-Producing and Integrating Hazard and Risk Knowledge. *International Journal of Disaster Risk Reduction*, 13, 242–247. https://doi.org/10.1016/j.ijdrr.2015.05.009.

Ferreira, T., Vicente, R., Mendes da Silva, J., Varum, H., Costa, A., and Maio, R. (2016). Urban Fire Risk: Evaluation and Emergency Planning. *Journal of Cultural Heritage*, 20, 739–745. https://doi.org/10.1016/j.culher.2016.01.011.

Forino, G., Mackee, J., and von Meding, J. (2016). A Proposed Assessment Index for Climate Change-Related Risk for Cultural Heritage Protection in Newcastle (Australia). *International Journal of Disaster Risk Reduction*, 19, 235–248. http://dx.doi.org/10.1016/j.ijdrr.2016.09.003.

Garcés Feliú, E. (1999). *Las ciudades del salitre: Un estudio de las oficinas salitreras en la región de Antofagasta*. Santiago: Impresos Esparza.

González Miranda, S. (2002). *Hombres y mujeres de la pampa: Tarapacá en el ciclo de expansión del salitre*. Santiago: Centro de Investigaciones Diago Barros Arana. Lom Ediciones.

Huang, D., Li, L., Zhang, H., Shi, L., Xu, C., Li, Y., and Yang, H. (2009, January). Recent Progresses in Research of Fire Protection on Historic Buildings. *Journal of Applied Fire Science*, 19, 63–81. http://dx.doi.org/10.2190/AF.19.1.d.

ICCROM and Government of Canada, Canadian Conservation Institute. (2016). *A Guide to Risk Management of Cultural Heritage, Sarja*. www.iccrom.org/publication/guide-risk-management.

Jigyasu, R. (2005, October 17–21). *Towards Developing Methodology for Integrated Risk Management of Cultural Heritage Sites and Their Settings*. 15th ICOMOS General Assembly and International Symposium: Monuments and Sites in Their Setting – Conserving Cultural Heritage in Changing Townscapes and Landscapes, Xi'an, China. http://openarchive.icomos.org/id/eprint/326/.

Jiménez, B., Pelà, L., and Hurtado, M. (2018). Building Survey Forms for Heterogeneous Urban Areas in Seismically Hazardous Zones: Application to the Historical Centre of Valparaíso, Chile. *International Journal of Architectural Heritage*, 12(7–8), 1076–1111. http://dx.doi.org/10.1080/15583058.2018.1503370.

Macuer Llaña, H. (1930). *Manual práctico de los trabajos en la Pampa Salitrera*. Valparaíso: Talleres Gráficos Salesianos.

Masi, A., Chiauzzi, L., Samela, C., Tosco, L., and Vona, M. (2014). Survey of Dwelling Buildings for Seismic Loss Assessment at Urban Scale: The Case Study of 18

Villages in Val d'Agri, Italy. *Environmental Engineering and Management Journal*, 13(2), 471–486. http://dx.doi.org/10.30638/eemj.2014.051.

Ravankhah, M., Chmutina, K., Schmidt, M., and Bosher, L.S. (2017). Integration of Cultural Heritage into Disaster Risk Management: Challenge or Opportunity for Increased Disaster Resilience. In: Albert, M.-T., Bandarin, F., and Pereira Roders, A. (Eds.), *Going Beyond – Perceptions of Sustainability in Heritage Studies*, vol. 2, no. 5. Cham: Springer. http://dx.doi.org/10.1007/978-3-319-57165-2_22.

Ravankhah, M., and Schmidt, M. (2014, September 8–10). *Developing Methodology of Disaster Risk Assessment for Cultural Heritage Sites*. Paper Presentation. Proceeding of Residential Doctoral School (RDS) at 4th International Conference on Building Resilience, Salford Quays, pp. 13–22. www.preventionweb.net/publications/view/41475.

Sesana, E., Gagnon, A., Bonazza, A., and Hughes, J. (2020). An Integrated Approach for Assessing the Vulnerability of World Heritage Sites to Climate Change Impacts. *Journal of Cultural Heritage*, 41, 211–224. https://doi.org/10.1016/j.culher.2019.06.013.

UNESCO, ICCROM, ICOMOS, and IUCN. (2010). *Managing Disaster Risks for World Heritage*. Paris. https://whc.unesco.org/en/managing-disaster-risks/.

UNESCO and Katholieke Universiteit Leuven. (2012). *Risk Management at Heritage Sites: A Case Study of the Petra World Heritage Site*. Jordan. https://unesdoc.unesco.org/ark:/48223/pf0000217107.

Yen, Y., Cheng, C., and Cheng, H. (2015). Disaster Risk Management and Measurement Indicators for Cultural Heritage in Taiwan. *ISPRS Annals of Photogrammetry, Remote Sensing and Spatial Information Sciences*, II–5(W3), 383–388. https://doi.org/10.5194/isprsannals-II-5-W3-383-2015.

11 Disaster Risk Management Plan for Punakha Dzong, Bhutan

Junko Mukai

Introduction

The Cultural Heritage bill of Bhutan 2016 stipulates that a district administration shall prepare a DRM plan for designated heritage buildings located under their jurisdiction. For preparing the first ever DRM plan of a heritage building in Bhutan, the Punakha Dzong, one of the most iconic Dzong architecturally and historically, was chosen. Recognising the importance of heritage buildings as a centre of district in DRM, World Bank provided support to prepare a DRM plan for Punakha Dzong.[1] It was executed in the framework of an ongoing project of building up Bhutan's resilience to disasters and climate in collaboration with the Department of Culture (DOC).

Firstly, the chapter defines Punakha Dzong's cultural heritage value and the disaster risks it faces to describe the project overview. Subsequently, it highlights the approach taken to develop the DRM plan. The chapter concludes by emphasising on the importance of continuous cooperation among the stakeholders.

Punakha Dzong – A Heritage Building of Special Importance

Punakha Dzong is one of the first to be designated a heritage building of special importance by the state. In Bhutan, the registration and designation system for built heritage commenced in 2021, and the DOC is currently listing up heritage buildings to register. Punakha Dzong is the masterpiece of Bhutanese architecture. Dzong originally means a fortified Buddhist monastery. Punakha Dzong stands on a hilltop like a gigantic vessel of massive high stone walls with minimum openings. The building complex was extended over time. The Dzong which we saw before the fire was completed by the late 18th century. The Dzong complex is spread over 200 m by 80 m. There are three courtyards inside the complex, two of the courtyards are 7 m high, and buildings and towers surrounding the courtyard create the complex's inner space.

DOI: 10.4324/9781003356479-14

Figure 11.1 Punakha Dzong.

Source: Author

Founded in 1637 by Zhabdrung Ngawang Namgyal, the prince abbot of Drukpa-Kagyu school – one of the Tibetan Buddhist schools and the unifier of the country – Punakha Dzong is the prominent centre of the country's culture and sociopolitical development. Punakha Dzong is the first stronghold where Bhutan's administrative system was formed. It was built as the principal seat of the founder, and later it was handed over to two successive leaders, Je Khenpo and Druk Desi, for spiritual and secular activities. Even today, the Dzong is habited by the Central Monk Body headed by Je Khenpo as their winter residence to upload the spiritual traditions. Currently, the Dzong accommodates the Central Monk Body and Punakha district administration offices, thereby sustaining the functional essence of housing spiritual and secular dual authorities. Owing to its historical significance, the Dzong has hosted important national events, including the royal coronation, wedding, and other important religious rituals.

The Punakha Dzong also shelters important Buddhist relics and artefacts, including those inherited for over 380 years since the Dzong's establishment. There are 1,500 items enshrined in 43 Buddhist shrines listed as movable cultural properties by DOC. The collection includes some of the most important and irreplaceable national treasures.

Disaster Risk

Even today, the Punakha Dzong stands strong due to the uninterrupted efforts of the ancestors for over 380 years to mitigate disaster risks and recover from disaster impacts. Although, floods, earthquakes, and fire consistently threaten the Dzong.

(1) Floods

History suggests that a prediction determined the Punakha Dzong location, so it was built at a place resembling an elephant's head. The Dzong stands on the end of a narrow stretch sloping down a northern mountain. This landscape resembling an elephant head is surrounded by two rivers, Mochu and Phochu. Both rivers merge to become the Punatshangchu flowing southwards to the Wangduephodrang district. The upper stream of both rivers is glacier lakes. A study (Mool & Wangda, 2001, p.67) identified that the Phochu basin consists of 549 lakes and the Mochhu basin has 380 lakes. A sudden outburst of lakes led to the Glacial Lake Outburst Floods (GLOFs) in the past. Although, there is no record of direct damage to the Dzong structure. In 1994, a flood of Phochu severely destroyed a small temple situated a few metres away from Punakha Dzong. After this flood, Phochu that was running close to the east side of the Dzong was diverted to the present location of merely 100 m away from the Dzong.

Since 2011, an early warning system has been installed in the Punakha-Wangduephodrang valley, considering the increasing probability of GLOF. As per an annual report of Bhutan's National Centre for Hydrology and Meteorology (2017, p. 6). Ten stations for automatic monitoring of water level and 18 sirens have been installed as of 2016 to allow the people along the river to evacuate within a few hours.

(2) Earthquakes

Bhutan is located in a seismic zone. Ancient literature and scientific record confirm that a severe earthquake rattled the Punakha region in 1713 and 1897 (Department of Culture, 2019, p. 26, table 4). There is no record about any damages to the Punakha Dzong caused by either earthquakes or other earthquakes recorded within or around the country.

Recently, two earthquakes occurred in 2009 and 2011 that affected Bhutan. Eastern Bhutan was the epicentre of the 2009 earthquake of M6.1 which damaged 5,456 buildings (Department of Disaster Management, 2009, p. 8). Sikkim–Nepal border was the epicentre of the 2011 earthquake of M6.9 which damaged 7,378 buildings (Department of Disaster Management, 2011, p. 8). Damages in the Punakha district were not very serious, and only minor damage to the Punakha Dzong was observed on the parapet wall of stone masonry added later. These earthquakes were an alarm to the government. To address this risk, the DOC undertook multiple research projects to improve the seismic resilience of traditional construction. The ongoing joint research project for a six-year term is to prepare guidelines for retrofitting an existing traditional building with resilient features in new constructions. The project implemented static loading tests and shaking table tests of the specimens including full-scale houses. The outputs of these researches would help reduce seismic risks to the Punakha Dzong.

(3) Fire

A fire in 1986 burnt down the south-west corner of the Punakha Dzong. Bhutan's precious collection of Buddhist scripts turned into ashes in this incident. The structural damage caused by this fire resulted in a major renovation which was completed in 2004. However, this was not the only fire that damaged Punakha Dzong. According to monastic literature, the Dzong had been damaged by fire at least eight times[1] in history, which led to some significant restorations.

The Punakha Dzong is comparatively well-equipped for fire operation as it is the winter residence of the Central Monk Body and houses the most important national treasure. Fire prevention facilities like 24-hour disposition of dedicated personnel, fire detectors, alarms, extinguishers, hydrants, and water reservoirs are stationed in the Dzong. However, if a fire similar to the Wangduephodrang Dzong were to happen, then fire control would become a challenging task.

The Dzong is highly vulnerable to fire because of its construction materials. Its inner façades are gorgeously framed with timberwork, such as projected windows, balconies, and rows of wooden columns. Layers of wooden roofing structures overwrapping one another help fire to spread quickly. Under such substantial wooden structures are stored over a thousand ancient Buddhist artefacts made of dried clay, cloth, paper or wood. The Dzong built as a fortified monastery is characterised by its defensive design of minimum accessibility. The massive stone walls of up to 20 m in height, narrow, and a limited number of entrances and openings are major hindrances for firefighting and rescue operations.

Nonetheless, fire is preventable by people. Multiple ways to mitigate fire risks are implementable with care and limited resources.

Preparation of the Punakha Dzong DRM Plan

(1) Background

Since 1985, DOC had administrated cultural heritage protection when its former organisation – the Special Commission for Cultural Affairs – was established through a Royal Decree. The initial scope of the establishment was to protect movable cultural properties such as Buddhist relics and artefacts. Later in 2003, the Division for Conservation of Heritage Sites (DCHS) was officially set up to protect the built heritage like Dzongs and temples. After the Constitution of Bhutan was promulgated in 2008, it has been urged to develop a legal framework to protect built heritage. In 2012, drafting the first-ever act on built heritage began. Later, the draft was merged with the present Movable Cultural Property Act of Bhutan 2005 and the draft of Intangible Cultural

Heritage bill. The three documents were compiled as the Cultural Heritage bill of Bhutan 2016. The bill mandates district administration offices to prepare DRM plans for important heritage buildings.

DOC chose Punakha Dzong for the first DRM plan preparation due to its national importance and popularity. It is expected to create awareness about DRM for cultural heritage to accelerate DRM plan preparation for other important heritage buildings. Furthermore, also it aims at establishing a standard format and process to prepare a DRM plan that can be referred by district administration offices for other nationally important cultural heritage sites.

(2) Outline

The project for developing the Punakha Dzong DRM plan commenced in August 2019. The DRM plan defines the roles and responsibilities of all relevant stakeholders, including district administration and police officials, ministries, the national volunteer group, and the Central Monk Body. It also clarifies communication and command mechanisms among stakeholders. These mechanisms are integrated with an existing legislative and institutional framework, especially with the Punakha District Disaster Management and Contingency Plan (DDMCP), which provides the overall DRM framework of the Punakha district. Punakha DDMCP was prepared as a mandate by the Disaster Management Act of Bhutan 2013 by the Punakha district administration under the technical guidance of the Department of Disaster Management. This Punakha DDMCP was endorsed in June 2019.

The Punakha Dzong DRM plan specifies measures for disaster risk mitigation, preparedness for response, and post-disaster heritage protection. The plan aims to ensure the safety of 400 monks living in the Dzong, dozens of district officials, and 120,000 tourists annually visiting the Dzong. Furthermore, the DRM plan attempts to protect the irreplaceable heritage building and 1,500 movable cultural properties housed in different places in the Dzong.

The project focused more on non-structural measures that are immediately implementable without extensive budget provisions, than infrastructural measures that shall be implemented in the mid- to long-term plans. Thus, an interactive process of preparing the DRM plan has been undertaken to enhance the collaborative engagement of various stakeholders by organising workshops, field surveys, table exercises, and discussions.

An action plan was developed to engage the stakeholders and monitor the implementation of the plan. The DOC is planning to organise a mock drill to test the feasibility of the proposed measures for evacuation and rescue. After testing, the plan may be revised accordingly.

DRM Plan for Implementation

Unfortunately, the plans prepared during a project are often shelved upon completion. Especially DRM plans since people are optimistic about its non-recurrence and find it unrealistic to take it seriously. Implementation of a DRM plan is hardly prioritised among other projects due to limited financial and human resources. Hence, this project focuses on developing a feasible and sustainable implantation plan.

(1) Interactive Participation of Stakeholders to Enhance Potential Capacity

Every past disaster has triggered the adoption of remedial measures for risk reduction of Punakha Dzong, such as river deviation, seismic resilience studies, and installation of flood warning, fire alarm, and firefighting systems. These measures need to be complemented with the capacity development of stakeholders. Since Bhutan is a small country, continuous investment in infrastructural development and maintenance is not easy. Hence, it is important to focus on effective and collaborative stakeholder engagement in the DRM measures, which can be implemented without excessive budgetary and technical resources.

Hence, from its first stage, the project focuses on the interactive participation of stakeholders to develop a sense of ownership to the plan and create an in-depth understanding of risks and possible disaster situations.

Effective ways were explored at every stage of preparing the DRM plan to enhance potential capacity and collaboration between stakeholders. A series of stakeholder workshops were organised to identify disaster risks and simulate disaster situations. There were discussions about existing rules on the use of electrical appliances as this poses a serious fire risk. Hindrances to evacuation and rescue were identified, along with proposing remedy measures. After these exercises, each stakeholder's roles and communication mechanisms were discussed and agreed upon. Punakha Dzong's DRM plan consists of outputs from these workshops.

(2) Engaging Monks Residing in the Dzong

Nearly 400 monks reside in the Dzong, which was built in a defensive design that makes disaster response challengeable. An emergency operation for tackling a fire may lead to panic and chaos. The monk's residential space could be moved out of the Dzong by constructing a new facility nearby. However, sustaining monastic function in Dzongs comprises an essential part of cultural heritage value and is commonly considered an important tradition for spiritual well-being.

The large population in the Dzong poses a risk. However, with proper delegation of roles and awareness of disaster risks, they can protect the Dzongs from probable disasters through fire risk mitigation, evacuation, and rescue of movable cultural properties.

A fire incident in the early summer of 2012 destroyed the Wangduephodrang Dzong – another important Dzong located 15 km downstream from the Punakha Dzong. During this fire incident, it was observed that a group of monks climbed atop the roof of a shrine and removed the wooden roofing shingles to prevent the fire from spreading to the shrine structure. Due to a lack of human workforce, they could not tackle the fire; however, their efforts were inspiring. If the people in Dzongs have an instinctive dedication to protecting the Dzong, increasing their capacity for DRM could be helpful.

The project engaged the monk representatives of the Central Monk Body. Their knowledge about the Dzong infrastructure enabled them to lead the field surveys and tabletop exercises. A crucial part of the DRM plan comprises the contributions of the monks. In collaboration with the district officials, the monks are developing rules on fire safety. For instance, there are certain electrical appliances of higher voltage that are generally prohibited to use in the Dzong, such as water boiler, rice cooker and electrical heater. These rules were reviewed from realistic perspectives to avoid any misuse of such appliances. Some spaces were designated where such electrical appliances are allowed to use for genuine need, and a superintendent for such a space will be appointed.

The Central Monk Body's role is vital and irreplaceable to rescue the movable cultural properties. Due to secrecy and security reasons, only some monks in significant positions know the details of movable cultural properties.

(3) Combining DRM Measures With Existing Practice

The workshop participants were encouraged to explore customary practices exercised in the Dzong, with which the DRM measures may be integrated. If we could integrate the DRM measures in the already existing practices, the implementation would be more easy and sustainable, instead of introducing a totally new routine of the DRM measures.

Interestingly, a monk participant proposed to keep a bucket of water at an altar where oil lamps are placed on a table. Every evening, offering water in cups on the table are collected in the bucket to discharge. Instead of discarding the water, the monk proposed to keep it overnight as a tool for fire safety.

Earlier, there was a tradition of night watch in the Dzong. In traditional night watch, one monk shouldered a long stick, and another hit it to make a sound. However, this tradition was abolished after police began patrolling the area and a modern fire alarm system was installed. Nevertheless, night watch exercised by residents could efficiently develop awareness. Thus, reviving this practice is included as an agenda for further discussion.

Similarly, one of the DRM measures proposed in the workshop was promptly integrated with a project proposal that had been pending for a long time. The workshop participants proposed to create one more exit after a rescue simulation. Initially, it seemed difficult to implement the proposal immediately due to necessary procedures. However, making an exit route is on track by combining it with the ramp proposal. The Monk Body requested constructing a ramp attached to the backside of the Dzong for carrying meals to the Dzong's congregation hall for religious rituals, but it wasn't approved. Nevertheless, the ramp proposal has been revived for the purpose of using it as an emergency exit. The documents of earlier proposal help firster implementation.

(4) Learning Lessons From the Previous Disaster

A disaster could be an unexpected opportunity to understand the gravity of disaster situations. It provides guidance to improve the existing risk mitigation measures and make them practical. For instance, this project reviewed the Wangduephodrang Dzong fire incident of 2012 to evaluate the measures to rescue movable cultural properties.

In Dzongs, important religious artefacts are locked in a steel safe for security reasons. Each of those safes is placed in different shrines in the Dzong. Typically, the shrine caretaker possesses the key to the safe. In case of the caretaker's absence, the risk of damage to these artefacts locked in the safe increases marginally.

The Wangduephodrang Dzong fire was an example of this situation. While the fire was spreading, safety boxes could not be opened in time to rescue the items inside. Then, a big man determinedly lifted the safe to a balcony and threw it on the ground outside the Dzong. Due to the landing impact, most of the artefacts inside broke. Reflecting upon this incident, it was concluded that the action undertaken was optimum at the moment. Damage is repairable, but

Figure 11.2 Wangduephodrang Dzong in fire in 2012.

Source: Punap Ugyen Wangchuk

once the safe is opened, there are higher chances of loss and theft in a time of chaos such as fire. Based on this observation, procedures to rescue movable cultural properties were deliberated, and the Central Monk Body is currently preparing necessary rescue operations.

Conclusion

The project infers from an earlier foundation and will be an addition to future development. One of such initiatives was to set up an office in charge of heritage management under the Central Monk Body to which the participant monk belonged. This is a good start to the sustainable implementation of the DRM plan. The Central Monk Body finally consented to set up this office after reviewing evidence that the cooperation between culture and monks benefited both heritage protection and monastic needs. For instance, the serial project to renovate five Dzongs implemented by the DOC in the tenth five-year plan (2008 to 2013) provided a monastic dining space for each Dzong under the project. These dining spaces materialised the introduction of the communal dining system initiated by the Central Monk Body. It was urged for disciplinary reasons, but it also reduces fire risks caused by the practice of each monk cooking meals using gas and electrical appliances in their residential room inside the Dzong.

DRM is not a one-time action but a continuous process. Hopefully, this plan enables the DOC together along with the stakeholders to enhance DRM and heritage protection.

Acknowledgement

The author would like to express her gratitude to the Department of Culture and all the stakeholders who participated in this project for their cooperation and inputs. Also, thank the World Bank for the valuable opportunity to take part in the DRM project in Bhutan. Special thanks go to Ms. Dechen Tshering, who led this Technical Assistance from the World Bank and a former participant of ITC 2010.

Note

1 The author, having worked in DOC from 2001 to 2016 as a conservation architect, was also working on a management plan for Punakha Dzong in 2017, and has been engaged in this project as a World Bank consultant (Phuntsho, 2014; Dorji, 1995).

References

Department of Culture, Royal Government of Bhutan. (2019). *Punakha Dzong Heritage Site Management Plan Part 1 Management Plan of Dzong and Conservation zone*.

Department of Disaster Management, Royal Government of Bhutan. (2009). *National Recovery and Re-Construction Plan: Building Back Better: September 21, 2009 Earthquake*. Department of Disaster Management, Royal Government of Bhutan. https://www.adrc.asia/documents/dm_information/Bhutan_National_Recovery_and_Reconstruction_Plan.pdf.

Department of Disaster Management, Royal Government of Bhutan. (2011). *National Recovery and Re-construction Plan: Building Back Better: September 18, 2011 Earthquake*. Department of Disaster Management, Royal Government of Bhutan. https://www.ddm.gov.bt/wp-content/uploads/downloads/acts&rules/National%20Recovery%20and%20Reconstruction%20Plan,%202011.pdf.

Dorji, C.T. (1995). *A Political and Religious History of Bhutan*. New Delhi: Prominent Publishers.

Mool, P.K., and Wangda, D. (2001). *Inventory of Glaciers, Glacial Lakes and Glacial Lake Outburst Floods: Monitoring and Early Warning Systems in the Hindu Kush-Himalayan Region – Bhutan*. Kathmandu: ICIMOD & UNEP.

National Centre for Hydrology and Meteorology, Royal Government of Bhutan. (2017). *Annual Report (July 2016– June 2017)*. National Centre for Hydrology and Meteorology, Royal Government of Bhutan.

Phuntsho, K. (2014). *The History of Bhutan*. London: Vintage Books.

Part 4

DRM Plan Implementation – Capacity Building/Others

12 From Theory to Practice

Insights From the Pathway to Implement DRM Measures for Cultural Heritage Sites

Monia Del Pinto

Introduction: Ideating a DRM Plan for Cultural Heritage Sites

The DRM project discussed in this chapter was ideated for the MuNDA, Museo Nazionale D'Abruzzo, Italy, during the 2019 ITC on Disaster Risk Management for Cultural Heritage Sites (R-DMUCH, 2020, pp. 50–56). The chosen museum is located in the city of L'Aquila, Italy, and is known in the Abruzzo region for its collections of archaeological remains, artefacts, art, and everyday objects, narrating local history and culture since prehistoric times. The museum was established in 1951 and is located within the Spanish fortress, a 16th-century building that is also a landmark of the city. The 2009 L'Aquila earthquake heavily damaged the fortress and part of the collections of the museum that were displayed in its spaces. In the aftermath of the 2009 earthquake, while the structure of the fortress was propped-up before undergoing restoration, all the collections were moved: the damaged ones underwent restorative measures, whereas the others were stored, awaiting the museum to reopen. After five years of inactivity in 2015, MuNDA was relocated to a different area in the city. The new location, in Borgo Rivera, was the repurposed historical slaughterhouse of the city, adjacent to the historical urban walls. The ITC Disaster Risk Management plan was ideated based on the museum's reopening in the permanent location of the Spanish fortress after the completion of the restoration work. In line with ICCROM guidelines (Tandon, 2018), the proposed DRM measures for the fortress were designed to replicate the 2009 earthquake scenario, combined with heavy rainfalls that have become increasingly frequent in the area during the last decade.

Proposed DRM Plan for MuNDA

The DRM plan was outlined after evaluating the heritage elements and values, and the risk assessment of the site, building, and collections. The heritage site consists of a pine grove park surrounding the fortress, the fortress building, and moveable elements of the collections. The site encompasses natural,

DOI: 10.4324/9781003356479-16

Table 12.1 Summary of hazard and vulnerability assessment in the MuNDA

Risk assessment	
Hazard	**Primary**: earthquake, rainfall **Secondary**: structural failure and collapse of walls and ceilings
Vulnerability	**Physical**: a. building typology (a fortress, designed not to be accessible, hence offering reduced evacuation routes and open spaces) b. building materials (heterogeneous materials used in previous restoration works, having different response to the seismic action) c. display choices in galleries and exhibition rooms (increased exposure of artefacts that were not encased, statues and paintings not adequately anchored to walls or ground) **Attitudinal**: a. lack of a multi-hazard emergency plan for the museum, and lack of a shared plan among the occupants of the structure – that is, superintendence, research institutes, and the MuNDA b. lack of knowledge and communication of emergency procedures

Source: Author

architectural, artistic, and historical values associated with each component. The hazard assessment focused on primary and secondary hazardous conditions featured in the scenario. Primary hazards encompassed earthquake and heavy rainfall, whereas secondary cascade hazards were the expulsion of debris due to structural failure and collapse of walls and ceilings. The vulnerability assessment looked at physical and attitudinal vulnerability: the former linked to the building typology, materials, and the layout of exhibition rooms and galleries, and the latter related to a lack of a shared, multi-hazard emergency plan (see Table 12.1).

Scenario

The earthquake scenario encompassed structural and non-structural failure of galleries, with the collapse of vaults and walls, and detachment of plaster and masonry, directly damaging the collections. Further damage was expected as a result of incorrect display measures, such as improper or lacking fastening for paintings and sculptures, and the absence of protective cases for the smaller elements. The fortress' structure – a medieval defensive construction accessible through a bridge and surrounded by a moat – offered narrow access and egress corridors that were blocked by debris, and the absence of alternative emergency exits made the rescue difficult.

Proposed Mitigation Measures

The project suggested structural and non-structural measures to be implemented at the site, building, and the collections to be displayed.

The proposed mitigation and preparedness measures were centred on avoiding, blocking, detecting, reducing the hazards, and building on existing capacities.

They included, but were not limited to, retrofit and consolidation of the structure, interventions on floors and shelving systems to mitigate the earthquake impact on objects; archive digitalisation and creation of heritage ID cards for the artefacts to minimise the loss of information and facilitate salvage (to locate and prioritise); improvement of the emergency plan including evacuation paths and redesign of exhibition areas; communication of emergency plan and coordination of actors (drills).

The recovery plan consisted of implementing response measures in the post-disaster phase, including recovery and relocation of the collection in safe temporary storage (Tandon, 2018, pp. 129–135). The plan was complemented with a series of proposals for business continuity, such as the museum's temporary relocation and launch of activities in partnership with other local, national, and international museums, and cultural NGOs. The activities were designed to maximise stakeholder engagement.[1]

Pilot Project, MuNDA: the Museum as a Catalyst for DRR4CH

Among the activities designed for the business continuity plan was the proposal for a pilot project to strengthen the museum's role in disaster risk communication and awareness, making the institution a local catalyst for Disaster Risk Reduction (DRR) measures for cultural heritage. In the project, lasting one semester, the museum plays a central role in educating pupils about heritage and DRR, with co-participation of the local primary and secondary schools, civil protection, fire brigades, and ITC resources. By involving the pupils' families through school activities, the project aims to enable cross-communication and facilitates broader community engagement.

Towards the Project Implementation

Interacting With the Institution

The DRM project for MuNDA was completed in October 2019, presented to the museum in December 2019, and reviewed between January and February 2020.

In February 2020, the museum became independent from the central directorate, gaining financial, scientific, and management autonomy.[2] The change implied, on the one hand, the prospect of a reduced bureaucracy – as a result of autonomy from the regional administration – and, on the other hand, the appointment of a new manager, with the consequent temporary interruption of the review process for the DRM project.

Between February and October 2020, the dialogue with the institution was on hold due to interim management change – with consequent delays on the implementation path.

Presentation and Review

In December 2019, the DRM project for MuNDA was first presented to the museum management team to discuss further customisation and possible review. The following sections describe the stages of presentation and review of the project, and insights on its pending implementation.

Step 1. Presentation

The informal meetings held in December 2019 and January 2020 in L'Aquila, Italy, welcomed the implementation of DRM plan in the permanent location of the Spanish fortress. The outgoing management was also willing to extend the plan to the temporary location of Borgo Rivera and to accordingly update the museum Service and Customer Charter after reviewing the project.

Consequently, the adoption of DRR measures for the temporary location was discussed. Special attention was given to strengthening the staff's preparedness, in order to adequately support response and evacuation activities in case of emergency. It was also suggested to broaden the scope of the pilot project '*MuNDA DRR4CH catalyst*' by involving a larger group of stakeholders – such as local municipality, academics, and professionals.

Step 2. Review – Suggestions, Additional Requests, and Customisation

In synergy with the management team, some of the original actions were prioritised and others were added *ex novo* to the original project (see Figure 12.1). The changes also implied a stakeholders' review (see Figure 12.2).

The revised actions were as follows:

a. Prioritise the staff training to enhance preparedness and internal capacity building
   This would imply assessing staff preparedness, and then designing and performing targeted training activities. Training would be delivered in one or multiple sessions, followed by periodic drills and updates.
b. Perform risk assessment on MuNDA temporary location, to inform the additional DRM plan
   This action would imply performing on-site survey in Borgo Rivera location, extended to building, galleries, and collections. The survey would be followed by hazard and vulnerability assessment for collections, structure,

and occupants, including staff and visitors, and the ideation of specific DRR measures.

c. Broaden the scope of the Pilot Project *MuNDA DRR4CH Catalyst*

The original version of the project was to be updated in line with MuNDA didactic section, so that activities could be implemented through the museum's teaching labs and delivered by pedagogists.

In its broadened version, the project would have encompassed training activities for higher education and professionals, in partnerships with the University of L'Aquila, professional orders, the Museum of XXI Century Arts (MAXXI), and local municipalities.

Step 3. Towards the Approval of the Revised Proposal

The DRM project for the MuNDA temporary location was left at the proposal stage due to management changes in combination with the disruptions caused by the COVID-19 pandemic that changed the priorities for the institution. The revisions scheduled in 2020 were not performed, and the dialogue with the institution was postponed to 2021. However, further management changes delayed the actions. In 2021, the DRM project appeared to be less and less likely to be implemented in the MuNDA temporary location, considering the imminent completion of the restoration works in the Spanish fortress – that is, the permanent location. However, the impossibility to establish a dialogue with the superintendency to incorporate DRR measures within the fortress' ongoing restoration plans suggests, in the best of cases, that the museum will have to approve and implement the DRM plan only after reopening in

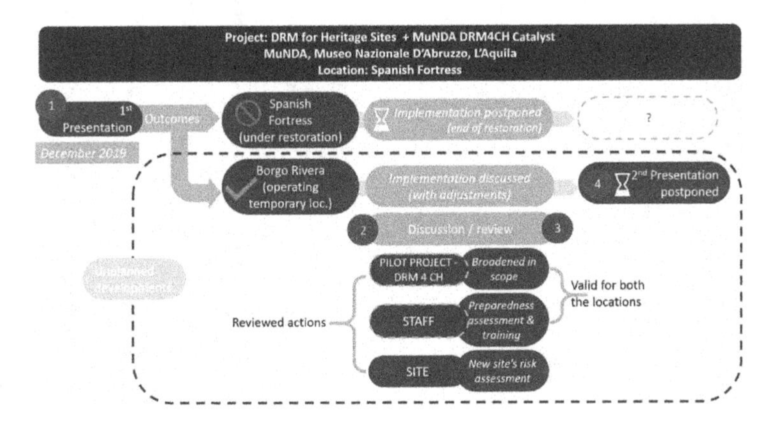

Figure 12.1 Summary of steps undertaken between December 2019 and August 2020.

Source: Author

Stakeholders in the reviewed project

Stakeholders	Role
- Museum (institution)	- Implements the DRM plan - Hosts training activities
- Museum staff - Citizens (visitors) - Research institutes AQ (universities, cultural institutes) - Schools (primary, secondary) - Professionals (architects, engineers, conservators)	- Receive training about preparedness, DRR 4 CH, CH salvage - Involved in lecture series
- Municipality of L'Aquila	- Can sponsor activities and receive visibility - Can receive training (DRR 4 CH)
- Fire brigades - Local civil protection (& volunteers) - Local police	- Support DRR-related activities and evacuation drills - Receive training about the CH salvage
- CH NGOs and other partners (DMUCH and ITC resources)	- Support and deliver training activities
- Donors (local/national/international)	- Sponsor the activities, receive visibility
- Ministry of CH and activities and tourism - Superintendence for Cultural Heritage (L'Aquila region) - Regional Secretariat of Abruzzo	- Can sponsor activities and receive visibility - Can receive training (DRR 4 CH)

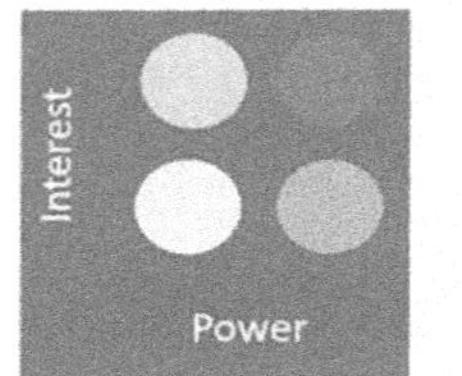

Figure 12.2 The stakeholders outlined for the reviewed project.

Source: Author

the Spanish fortress. The plan, however, would probably require additional review to fit the building and its surrounding after restoration.

Discussion and Key Learning Points

Since the implementation was stalled, it was not possible to assess if, and to what extent, the DRM plan for MuNDA and the pilot project were effective in strengthening DRR capacity within and outside the institution. However, some reflections were prompted in the ten months following the project's ideation, that could be looked at as learning points for practitioners.

The following points summarise the recommendations to inform DRM action at several stages:

1. Enable Flexibility of the DRM Project for Cultural Heritage Sites

A DRM project for a cultural heritage site is a living document subject to review, improvement, and constant customisation. Adapting the project – at any stage – may deviate from the original purpose but shows the only possible way to meet the need of the institution and the immediate context. Working in synergy with the recipient institution through consultation and feedback loop will ensure specificity of measures and maximise their success when implemented.

2. Recipient Feedback Is as Important as Consultation

As shown in the case of MuNDA, the museum management – specifically, the former director and the collection curators – played a crucial role in indicating specific priorities and informing the review of the DRM plan. Thanks to their roles within the institution, they provided insights into the areas that required urgent improvements and suggested how to shuffle priorities in the DRM plan. An example is offered by the request for targeted preparatory training with the museum staff – at all levels – prior to any other initiative, to fill the internal lack of DRR knowledge applied to the museum context.

3. Providing for Bureaucratic Contingencies

When operating in post-disaster contexts, or in the case of museums managed by the State, bureaucracy can become an obstacle in that generating delays and path changes – an aspect that should be factored in when a DRM for CH project is first ideated.

In the case of MuNDA, the months of inactivity following the project presentation have shown the delays due to the chain of approval within the Italian Cultural Heritage institutional structures (General Directory, Regional and Local Museum Centres). The detrimental role of bureaucracy in slowing down initiatives, causing postponements and delays, can have multiple impacts – the first of which is the possible need for further reviews and adaptations once the project is finally resumed. This would be necessary, especially in case of long delays, to avoid the risk of measures that might not fit the mutated needs of the specific cultural heritage site. An additional impact of postponements is the possible loss of momentum, budget, as well as stakeholders' interests. Lastly, delays in DRR action can directly affect the disaster risk for the site, in that, in the absence of DRR measures, it is exposed to the impact of possible disruptions – as shown by the COVID-19[3] emergency that happened while the museum had no mitigating measures in place.

4. Prioritise Low-Budget Activities

When a DRM for CH project remains on hold, it is challenging to evaluate the impact that future available budget will have on the planned measures. Low-budget activities and non-structural measures should be prioritised to minimise further setbacks and delays. This would also establish an initial virtuous framework to build upon once resources become available.

Conclusion

This chapter has presented the non-linear pathway to review and implement a Disaster Risk Management project for heritage sites ideated within the themed 2019 UNESCO Chair International Training Course. The case has provided one emblematic example of disconnect between the expected implementation pathway and the actual barriers that can be recorded on the ground. The process provided valuable learning opportunities and generated reflections on challenges posed to the DRR professionals committed to the integrated protection of cultural heritage. The reported experience highlighted that acquiring skills to ideate a DRM project for a cultural heritage site and working in synergy with highly interested stakeholders is necessary but not always sufficient towards a successful implementation of DRR measures. In specific cases, such as the presented post-earthquake context, the highly bureaucratised, state-driven restoration of heritage sites can pose barriers to multi-stakeholders' DRR action. Similarly, the discontinuity in management of a cultural heritage institution, such as the presented museum, could cause reallocation of resources – both budget and staff – or changes in strategies. It appears clear that, besides specific DRR knowledge, DRM for CH professionals are required the ability to minimise the impact of delays, modifications, and budget restrictions on the original project. In training cultural heritage professionals, it is, therefore, crucial to foresee the challenges that may arise in the implementation stage. Anticipating the possible, context-specific obstacles to

DRM for CH action would be key to inform strategies to overcome barriers on the ground. It is also important to consider that the number of adaptations and trade-offs needed for a project to be implemented represent valid indicators of i) alignment of the proposed DRM measures to the contextual needs and ii) disposition and sensibility to DRM measures from local stakeholders. These aspects, in turn, are the starting point to inform synergic action. Lastly, only a part of the barriers can be overcome by an individual effort. The cultural barriers and lack of disaster risk culture that slows down the projects must be tackled collectively on multiple scales, from theory to practice in disaster risk management, and in the cultural heritage sector.

Notes

1 Promotional activities, involving cultural NGOs (ICCROM), Civil Protection, and the stakeholders that were previously sharing the location with MuNDA (Concert Society of L'Aquila, INGV, and Research Institutes).

- Partnership with the local university
- Creation of a network of museums

2 DPCM, December 2, 2019, n. 169, on the reform of cultural heritage.
3 Having specific DRR measures in place (such as a digital archive, or DRR4CH initiatives scheduled) during the COVID-19 lockdown, would have facilitated the virtual/remote activities of the museum and supported the business continuity (e.g. in support of didactic activities).

References

Del Pinto, M. (2020). *Disaster Risk Management Plan for Heritage Sites: A Proposal for MuNDA*. L'Aquila, Italy: Museo Nazionale D'Abruzzo.

D-MUCH. (2020). *International Training Course (ITC) on Disaster Risk Management of Cultural Heritage*. Paper Presentation. Proceedings of XIV UNESCO Chair Programme on Cultural Heritage and Risk Management. https://rdmuch-itc.com/wp-content/uploads/Proceedings-of-ITC2021-2.pdf.

Dvorak, J., Burkšienė, V., and Sadauskaitė, L. (2019, June). Issues in the Implementation of Cultural Heritage Projects in Lithuania: The Case of the Klaipeda Region. *Cultural Management: Science and Education*. https://doi.org/10.30819/cmse.3-1.02.

Fatoric, S., and Seekamp, E. (2017). Securing the Future of Cultural Heritage by Identifying Barriers to and Strategizing Solutions for Preservation Under Changing Climate Conditions. *Sustainability*, 9. https://doi.org/10.3390/su9112143.

ICCROM. (2021). *Heritage in Times of COVID*. www.iccrom.org/first-aid-collections.

MIBACT. *Qual è il ruolo della Direzione generale Musei? Quali sono le sue funzioni e come è organizzata?* http://musei.beniculturali.it/struttura

Presidenza del consiglio dei Ministri. (2019, December 2). *Regolamento di organizzazione del Ministero per i beni e le attivita' culturali e per il turismo, degli uffici di diretta collaborazione del Ministro e dell'Organismo indipendente di valutazione della performance*. DPCM n. 169. https://www.beniculturali.it/provvedimenti-in-evidenza.

Tandon, A. (2018). *First Aid to Cultural Heritage in Times of Crisis*. Rome: ICCROM.

13 Earthquake and Cultural Heritage

A Rescue Project in Central Italy

Barbara Caranza

Introduction

In 2013, CHIEF ETS, a voluntary civil protection association, was established to safeguard cultural heritage in a crisis situation. Professional restorers and qualified restoration technicians are members of this network. The association also includes biologists and chemists specialising in cultural heritage, architects, engineers, and archaeologists. In 2015, it became the first Italian association to be included by the International Blue Shield in the list of six organisations in the world that, alongside its national committees, are recognised for the protection of cultural heritage in an emergency situation.

It is the first civil protection voluntary association in the Italian panorama constituted of professionals that have intervened in the emergency area upon request of the Ministry of Cultural Heritage and Tourism (MiBACT) of the National Department of Civil Protection and the municipalities affected by the impact of natural events of different origins.

Additionally, new volunteering groups were recently established within other civil protection associations dedicated to cultural heritage rescue. Since these groups do not have qualified professionals, their employment possibilities and potential are severely limited.

The training of cultural heritage professionals involved as volunteers in the activities of CHIEF ETS includes in-house training courses that aim to provide participants with knowledge of the civil protection system, its *modus operandi*, and the legislative framework of different operative scenarios.

The training courses organised within the association also provide tools to manage stress and teamwork, preparing them to work in a harsh environment and effectively interact with the affected communities.

The relationship between the rescue activities of the volunteers and the affected communities is a crucial issue for the association. Each intervention's purpose is first to save the property for its inherent values and stabilise and secure it. However, the ultimate focus of every intervention is the community.

DOI: 10.4324/9781003356479-17

The cultural heritage is saved first due to the importance of its intangible values for the community, and because it has the potential to transform itself. Through the long process of recognition and recovery along with the community, unexpectedly, the good acquires new values wherever resilience is needed.

The specialised knowledge of the teams enabled them to lead all the operations with responses calibrated to the type of cultural property affected and the disaster.

The association has specialised resources and equipment for immediate intervention after the impact. For instance, if water or mud has damaged paper or photographic artefacts, the readiness to react is an essential factor for the success of the intervention. Therefore, there is a strong need for defined and tested protocols. Since emergency procedures cannot be improvised, the success of every intervention is the result of theory and practice. It is mandatory to plan the intervention during the emergency phase that does not concern the building itself.

The activities undertaken by the association include sample collection for scientific analysis, disinfection using gamma rays, recovering, packaging, and cataloguing by filling out specific charts proposed by both the civil protection system and the MiBACT. Moreover, it includes stabilisation interventions, monitoring of thermo-hygrometric environmental values and assets, handling,

Figure 13.1 The CHIEF ETS logo.

Source: Author

and in the case of paper or wooden assets, freezing them at suitable dedicated facilities owned by members of the association.

CHIEF ETS firmly believes that the practice is a fundamental component of the success of all emergency operations. Therefore, it wants to record them as publications, conferences, and workshops. In addition, the association evaluates errors, processes them, and constitutes the starting point for future new interventions.

Researching and studying new methodologies for stabilisation operations on movable works and containers is fundamental for the association.

2017 Earthquake in Central Italy – Intervention by CHIEF ETS at San Salvatore, Campi (Norcia)

In 2016, following the seismic sequences that affected Norcia, the church of San Salvatore, built in Campi, completely collapsed.

San Salvatore, formerly Pieve di Santa Maria, stands in the old city-centre area of Campi in Norcia. The left aisle has a triumphal frescoed arch. The arch frames a monumental iconostasis with an upper attic. All the paintings in the lower area of the iconostasis date back to 1466. They are attributed to Nicola da Siena, signed with a date under a fragment of the enthroned Madonna on the left.

The purpose of the activation was to provide professional volunteers qualified to proceed with the rescue, cataloguing, stabilising, and handling operation of the frescoes and stone elements fragments from the Basilica of San Benedetto in Norcia and the Church of San Salvatore in Campi. The Coordination of the UCC of Umbria decided to avail CHIEF ETS exclusively for San Salvatore since the Basilica of San Benedetto in Norcia was considered structurally unsafe.

Before accepting the assignment, CHIEF ETS undertook surveys to evaluate the safety measures of both the sites for planning the intervention, scheduling the necessary human and instrumental resources, and evaluating the decrease in seismic activity.

The work was completed in eight weeks, where the teams worked for six days, starting Sunday and ending on the following Saturday. Eight volunteers were involved, and there was provision for more volunteers to work in shifts. Superintendence of Archaeology, Fine Arts and Landscape of Umbria provided Boarding and lodging at the dei Cacciatori hotel in Biselli, Norcia.

Each volunteer was officially enrolled to CHIEF ETS and equipped with safety shoes, knee pads, helmet with a chin strap, gloves, a dust mask, and goggles.

The work led by CHIEF ETS included excavation and selection of 'A' category rubbles, stabilisation, cataloguing, and handling of US1 (GREEN), US2 (GREEN), US3 (GREEN), A, B, C, D, US1 (RED), US2 (RED), US3 (RED), US4 (RED) sectors.

The Central Institute of Restoration (ISCR) decided on the collapse maps and catalogue systems which the CHIEF ETS operators accurately led. The intensive eight weeks of co-working of these trained eight teams allowed the recovery, stabilisation, cataloguing, and handling of 65% of the total collapsed church area. The handling of selected, stabilised, and catalogued assets was done according to the ISCR, the Superintendence of Archaeology, Fine Arts and Landscape of Umbria, and the Cultural Heritage Protection nucleus of the Carabinieri of Perugia. The operation commenced from the Church of San Salvatore to the temporary deposit of Santo Chiodo, located in Spoleto, with the assistance of vehicles owned by the association's volunteers.

The frescoes inside the building collapsed along with the walls, and they showed relevant problems of cohesion and adhesion to the stone ashlars that originally served as supports. In addition, several phenomena of de-cohesion and de-adhesion regarding the residual pictorial layers and accumulations of deposit materials were found.

Deferred damages due to meteorological events such as chemical-physical degradation and thermo-hygrometric parameters incompatibility were also identified. The frescoes fragments were stabilised by adhesion between the plaster layer and the paint film, using acrylic resin in primal aqueous emulsion, applied by injection or brush near the detachments.

Japanese paper, gauzes, and paraloid resin were used for all the cracks and detachments, and ethafoam wedges were placed in the cracks.

The fragments of the pictorial film were placed, divided by sector, in boxes covered with neutral pH tissue paper. After being catalogued, they were

Figure 13.2 San Salvatore a Campi – the volunteers of CHIEF ETS during the rescue phases.

Source: Author

Figure 13.3 San Salvatore a Campi – detail of the operations of consolidation of the paint film.

Source: Author

transferred to the temporary storage of Santo Chiodo along with the other discovered movable works.

CHIEF ETS and the Frascaro Community

During the rescue operation, volunteers met the Frascaro community, a small village near Norcia. Norcia is an Italian town of 4569 inhabitants in Perugia province, Umbria, Central Italy. Historically rich in testimonies, Norcia bears traces of human settlements dating back to the Neolithic (VI-V millennium BC). The Romans conquered the city at the beginning of the 3rd century BC. Norcia became one of the most important centres in the Italian Middle Ages (14th century). Norcia is a triumph of architecture and refined art history, but it is constantly threatened by earthquakes and catastrophic events strongly linked to the geomorphology of the place.

CHIEF Onlus had not received any assignment to work in Frascaro. The Cultural Property Protection Company of the Italian Army led all the operations in Frascaro, such as the recovery of the mobile works buried in the collapsed Church of Sant'Antonio.

The community's devotion to Saint Patron and the Church increased after the collapse. For people, the need to regain contact with their body and environment and not be detached from the trauma scape is the first step to consider for decreasing the victim's awkwardness and raising the feeling of belonging to that particular community. Accordingly, the cultural property can become a place where people can build resilience and imbibe new values. To protect cultural goods, the community must be involved in the practical work and the decision-making process to allow them to recover from the tragedy positively. This concept is still very underdeveloped by both the institutions and the rescuers themselves as they identify humanitarian aid mainly as the preparation of makeshift camps. These camps are good shelters for the short term, but in the long term, they can turn into isolated places where the victims lose a bond with their native land.

The Frascaro community now lives inside temporary containers and has recently founded a non-profit association to raise money for the reconstruction of the Church and restoration of all its artworks.

The community seemed frozen, awaiting change. They were already away from media spotlight, and all the reconstruction and stabilisation activities in that area were still at a standstill.

The Docufilm

The traumatisation process phases must be considered to understand the community involvement in the post-disaster phases in depth. Emergency psychology requires avoiding apathy. The natural tendency of people involved in crisis situations is to isolate themselves, which is harmful. Talking, networking, and sharing emotions and feelings about their personal story are imperative.

Since volunteers were not trained for psychological therapies, they could not seek explanations about the trauma or interpret victim behaviour. Incorrectly sharing a trauma story can compound the problem.

The film director must consider these aspects. Hence, he proposed a silent movie with images and people speaking voluntarily.

The documentary became a game and work in which the community could recognise itself. As per the psychology of emergencies, this process helps victims regain lost contact with the changing environment during the post-trauma phase.

Alessandro Leone, a skilled film director, screenwriter, and member of CHIEF ETS, led the making of the docufilm 'Story of Stones' with sensitivity and delicacy.

The film director, a Special Silver Ribbon awardee of the 31st Turin Film Festival, has won numerous other awards in Italy and abroad. At the 67th Trento Film Festival, 'Stories of Stones' was awarded the RAI Award for the best documentary. The short film was included in the official selection of Torino Cinemambiente and was chosen among eight Italian documentaries screened at Cardiff Italian Film Festival.

The docufilm intertwines places and stories connected by the earthquake. The protagonists of the story are CHIEF and its volunteers, but the film aims to present a tribute to a culture and a community that strongly wants to survive and overcome the post-earthquake isolation.

Conclusion

In conclusion, the CHIEF ETS represents an immediate, practical, organised, and respectful solution for safeguarding cultural assets at risk. The presence of specially trained professionals to act in crisis environment allows for the success of the operations. After all, dealing with crisis environments requires timely, well-reasoned action to be conducted by professionals, not improvised volunteers, to cope with all possible difficulties. Effective response and contribution can only be made through such experienced intervention.

14 Future Approach

Rohit Jigyasu, Dowon Kim, and Lata Shakya

It is imperative to reflect on the key emerging themes based on various case studies showcasing various practices in disaster risk management of cultural heritage sites and institutions. Broadly, the chapters in this book lead to three general but interconnected inferences that are critical for implementing DRM plans. Firstly, acknowledgement of disaster risk management is a continuous process that involves maintaining a link with existing systems and learning from previous disasters. The first step requires using participatory tools to map both cultural heritage and the risks and vulnerabilities. This will necessitate the engagement of all stakeholders, including the local community, experts, policymakers, and individuals associated with the heritage site or institution. Other key themes emerging from various chapters are institutional collaboration, the role of traditional knowledge, and innovative technology for disaster risk reduction of cultural heritage. The following paragraphs briefly discuss these emerging themes that can contribute to developing future strategies.

Risk Assessment

Risk assessment is an important step for initiating a disaster risk management plan. Different typologies of hazards have been considered for assessment across chapters of the book. However, all case studies commonly demonstrate that non-linear paths grounded in the local context are the most effective way forward. Pinto, in her chapter, showcases the disconnect between the assumed linear implementation and the non-linear paths during the implementation phase. Grimaldi and Vargas adopted an innovative methodology of documentation of decorative elements of the built heritage. The decorative elements, a key component of the architectural vocabulary, contribute towards the significance of the historic building. While Hutardo's chapter discusses risk assessment for wind hazards, Ravankhah et al. describe the development of the cultural heritage risk index. The proposed tools provide a shared understanding of hazards and risks among multiple stakeholders engaged in

DOI: 10.4324/9781003356479-18

protecting heritage sites. The examples also highlight the importance of integrating structural and non-structural vulnerabilities of different typologies of cultural heritage at risk while addressing the spatial and temporal dynamics of vulnerability. Giuliani's chapter on the Tuscan site showcases the criticality of material knowledge and understanding of technology for hazard assessment. The cases advocate for a multi-hazard approach to risk assessment, as most heritage sites and institutions are likely to be affected by more than one hazard.

Stakeholder Engagement

The importance of stakeholder engagement also emerged as an important theme from various DRM initiatives. Grimaldi and Vargas have discussed the use of risk mapping to raise stakeholder awareness as the first step towards securing engagement. Mukai emphasises the value of interactive stakeholder participation to maximise their potential, fostering a sense of ownership and motivating them to take proactive measures. Tanner emphasised the need for using and maintaining existing relationships as part of an effective emergency management system. Saleh progressed further in his chapter, where the project included identifying the stakeholders and assessing their financial capabilities and technical recourses. George Town World Heritage Incorporated (GTWHI) has been a success story of community-centred approaches. Utilisation of a social network of local agencies and communities and involving them in complex activities can lead to good practice of community-based response strategy. The activities could be scientific analysis for risk assessment, inventorying, and setting up a geographic information system, monitoring system, and awareness campaigns. Caranza elaborates on the works of CHIEF ETS, a voluntary civil protection association focused on safeguarding cultural heritage in a crisis situation. The example of the historic village of Frascaro shows how such organisations can play an important role in empowering the local community in decision-making and practical work related to emergency response and recovery of cultural heritage.

Institutional Coordination

The involvement of various stakeholders, including institutions from cultural heritage and disaster risk management, is critical for managing disasters at multiple stages. Considering the need for timely execution, communication and seamless coordination among various institutions will be determining factors for the success of the management plans. Hurtado highlights the collaboration between academic institutions – the Technical University Federico Santa María, the national office in charge of World Heritage Sites, and the Site Managers. It was critical since the academic institution researched the risks involved, and the National Centre of World Heritages Sites (CNSPM) took responsibility for developing tools to reinforce the conservation and maintenance of the concerned sites. The case study is an excellent example of the need for academic institutions and agencies at national and site levels to work collaboratively. Tanner focused on the importance of using the language and framework established by New Zealand's National Emergency Management Agency to communicate and align the process developed for heritage sites. The example emphasised the importance of establishing and maintaining relationships with various institutions and stakeholders for effective communication at various stages. Castro's case example shows collaboration between the National Institute of Anthropology and History (INAH), the local government, volunteers, the community, and the local heritage conservation sector. Ravankhah et al.'s example also highlights the importance of the involvement of a multidisciplinary team of specialists and site managers.

Capacity Building

To effectively implement DRM plans, stakeholders must be empowered with the appropriate capacities. Hence, capacity building in the pre-disaster stage is vital, as Ang, Caranza, and Torres emphasised. Caranza discusses the courses for heritage professionals aimed at knowledge related to the civil protection system, its modus operandi, and the legislative framework for different disaster scenarios. In the case of George Town World Heritage city, the local institution, GTWHI, utilised expert knowledge from resource persons through the social network of local agencies and communities to involve them in various disaster management activities.

Traditional Knowledge

The 'Declaration', adopted at the International Disaster Reduction Conference (IDRC) of Davos in August 2006, confirmed that 'concern for heritage, both tangible and intangible, should be incorporated into disaster risk reduction strategies and plans, which are strengthened through attention to cultural attributes and traditional knowledge'. This was exemplified by the case study at Gjirokastra, where traditional rainwater collection has been transformed into an effective fire prevention system.

Innovative Technology

Being updated with new technologies and engaging with them is an important part of the management process. Ravankhah et al.'s case study exhibits implementing a web interface to assist domain experts and site managers in providing hazard and site-specific risk mitigation strategies and efficiently responding during an emergency. The process of using a pumping system in traditional water cisterns is an excellent example of innovative thinking showcased by Mamani's case study. Giuliani's example uses a Geographical Information System (GIS) based platform to create a decision support system for the Tuscan town walls based on the multidisciplinary and cross-scalar knowledge of all regional assets.

Index

ABC powder, condition (pressure check) 71–72
action plans: implementation 68; set 20
AED training 70
al-Azhar/al-Ghūrī district: applied method, outcomes 33–34; applied method, processes 33; built heritage, overall condition 27; case study 24; characterisation 25; conceptual framework 25; condition assessment 26; electricity boxes/telephone lines, location 34; fire hazard 29–31; fire hydrants/hoses, location 33; fire incidents, history 29; fire primary hazard, mitigation measures *30*; fire response/mitigation strategies, inadequacy 30–31; fire risk assessment (FRA) methodology, proposal 26, *27*; ignition, source 29–31; mid-term strategy, proposal (measures) 32; old buildings, demolition *28*; risk assessment 28–29; short-term strategy, proposal (steps) 32; significance 24–25; site, conservation (status) 27; streets, fixed/temporary obstacles (determination) 33; study zone 27; subject proposition 25; vulnerability 26; zone, accessibility (absence) 31
ancient town walls: city preservation *38*; failures, examples *41*; future developments 43; Pisa, town walls 41–43
ancient town walls, risks 37
ArcGIS mapping 46
archaeological decorative elements project 59
Argyriou, Athanasios V. 45
as low as reasonably practicable (ALARP) principle, application 52
Atfit el-Sokary, fire 29
Atfit Hosh el-Nimr, old buildings (demolition) *28*
Atfit Hosh Nimer, fire 29
Australia New Zealand Joint Scientific Committee on Risk Preparedness (JSC-ANZCORP) 13

Basilica of San Benedetto (Norcia) 130
Baths of Diocletion (Italy) 56
Bhutan: Constitution, promulgation 110; seismic zone location 109
Birkmann, Joern 45
buffer zones 18
building collapse, identification 88

Cantebury Earthquake Recovery Agency, recovery responsibility 12
capacity building 6–7, 137; action, implementation 59; community-based capacity building 69; impact 64; importance 72; staff training, prioritisation 122
'Capacity Building for Disaster Risk Management of Cultural Heritage: Challenges and Opportunities in Post-COVID Times' 5
'Capacity Building for Disaster Risk Reduction of Heritage Cities in Southeast Asia and Small Island Developing States in the Pacific' 66, 67

capacity-building tools, recognition 89
capacity building/training (Dominican Convent of Santo Domingo Tehuantepec) 89–90
Caranza, Barbara 128
catastrophic, matrix term (usage) 20
Central Institute of Restoration (ISCR) decisions 131
Central Italy earthquake: CHIEF ETS intervention (San Salvatore, Campi) 130–132; rescue project 128
Central Monk Body 108, 113–114
Chee Ang, Ming 66
Chliaoutakis, Angelos 45
Cholula (Mexico), decorative elements (risk/priority map) *63*
Christ Church Cathedral, earthquake damage 11, *12*
Christchurch earthquakes, heritage (loss) 11
Church of San Salvatore (Campi) 130
cisterns, selection 77
City Council of Penang Island, closed-circuit television cameras (installation) 70
Civil Defence and Emergency Management (CDEM) 14
Civil Defence Department, Ministry of Tourism and Antiquities (collaboration) 34
Civil Defence operating procedures, emergency planning 11
civil protection volunteering 128
climate change adaptation plan 52
climate hazard assessment 48
collapse maps/catalogue systems, Central Institute of Restoration (ISCR) decisions 131
combustible materials, contact (limitation) 32
commercial activities 31
commercial businesses: fire risk, increase 31; fire source, determination 34; monitoring 34
community-based activities, usage 6–7
community-based capacity building 69
community-based Disaster Risk Management 72–73
'Community-Based Disaster Risk Reduction Coordinating Workshop for George Town UNESCO World Heritage Site' 69
community-based fire preparedness/response strategy 71–72
Community-Based Fire Responders, inauguration 71–72
community engagement 6–7
coping capacities 49, 51
cost-effective analysis (CEA) 52
COVID-19 pandemic: fieldwork challenges 5; impact, Waitangi National Trust awareness 15; restrictions 64
critical threats, Humberstone and Santa Laura Saltpetre Works (Chile) 98–99
cultural heritage: attributes 88; Disaster Risk Management 2–3, 66–68; elements, enhancement 68; first aid 72; intangible cultural heritage 88; inventory list, backup copies 69; rescue, civil protection volunteering 128; risk index 45; risk mapping 59; shared asset 73; typologies, structural/non-structural vulnerabilities (integration) 136
Cultural Heritage, Disaster Risk Management (International Training Course) 1
Cultural Heritage International Emergency Force (CHIEF ETS) 128, 136; collapse maps/catalogue systems, Central Institute of Restoration (ISCR) decisions 131; cultural heritage, saving 129; deferred damages, identification 131; docufilm, usage 133–134; Frascaro community, interaction 132–133; intervention 130–132; intervention, success (factors) 129; logo *129*; paint film, consolidation *132*; volunteers, actions *131*; volunteers (rescue activities), communities (relationship) 128–129
cultural heritage sites: characteristics 45; Disaster Risk Management (DRM, research 2; Disaster Risk Management (DRM) measures, implementation 119; Disaster Risk Management Plan (DRMP), ideating 119; Disaster Risk Management Plan (DRMP)

project, flexibility (enabling) 125; mitigation measures, proposal 120–121; scenario 120
Cultural Heritage without Borders (CHwB) 74, 77; ChwB Albania, project concept development 78; team, meetings organisation 79
cultural significance, survey (Dominican Convent of Santo Domingo Tehuantepec) 86
cyclones, Meteorological Service (Mexico) registration 84

damage assessment, first aid 72
decorated elements, hazard risk 64
decorative elements: consideration/risk 60; documentation, preparation 61; negative impact 62
Del Pinto, Monia 119
Deuk Desi 108
De Wit, Rosmarie 45
Digital Elevation Model (DEM), input data set 48
Disaster Imagination Game tool, usage *87*
Disaster Management Responding Agency 68
disaster risk (Punakha Dzong) 108–110
disaster risk cycle *17*
Disaster Risk Management (DRM) 1, 66; cultural heritage elements, enhancement 68; cycle *3*; framework 16; implementation, community-based activities 6; research 2; theme 2–3
Disaster Risk Management for Cultural Heritage Sites (D-MUCH) 119
'Disaster Risk Management of Cultural Heritage: Learning from Japanese Experience' 5
Disaster Risk Management on Cultural Heritage posters, GTWHI printing 70
Disaster Risk Management Plan (DRMP): Historic Cairo need 24; Humberstone and Santa Laura Saltpetre Works 95; development, methodology *97*; ideation (MuNDA) 119; implementation, community-based activities (usage) 6–7; Punakha Dzong (Bhutan) 107; Punakha Dzong (Bhutan), outline 111; Punakha Dzong (Bhutan), preparation 110–111
Disaster Risk Reduction (DRR): activities, proposal *102*; geographic information system (George Town) 69–70; Sendai Framework 1, 66; stakeholder articulation proposal *103*
Disaster Risk Reduction for Cultural Heritage (DRR4CH): MuNDA pilot project 121; Pilot Project MuNDA DRR4CH catalyst, scope (broadening) 123
Disaster Risk Reduction Programme, efforts/challenges/gaps 67
Division for Conservation of Heritage Sites (DCHS), setup 110
docufilm 134
Dominican Convent of Santo Domingo Tehuantepec: capacity building/training 89–90; cultural heritage attributes 88; cultural significance, survey 86; data processing 87; Disaster Imagination Game tool, usage *87*; intangible cultural heritage 88; local cultural heritage, disaster risk management plan 84; local society, psychosocial strengthening (facilitation) 85; materials/methods 85–87; methodology/scope 86; objectives/goals 85; participatory risk mapping workshop 86–87; preparatory work 86; results/discussion 88–90; stakeholder analysis 85–86; vulnerability mapping 88; vulnerability/risk maps, generation *89*
Drukpa-Kagyu school 108

earthquake risk (Punakha Dzong) 109
'Earthquake Risk Management of Historic Urban Areas' 4
earthquakes: National Seismic Service database (Mexico) 84; risk index, development *51*
electrical needs, rearrangement 32
electrical transformers, movement 32
electrical wiring, upgrading 32
electricity boxes/telephone lines, location 34
emergency evacuation, planning 2–3

emergency response, ITC focus 4
emergency response plan 52
Emergency Response Team: GTWHI setup 70–71; Training Sessions 71
Ephesus 45
'Exemplary Practice Award' 6
exposure assessment 46, 49

fire: community-based fire preparedness/response strategy 71–72; hazard 29–31; hydrants (Gjirokastra, Albania) *79*; hydrants/hoses, location 33; ignitions 29–30; ignition source 33; incidents, history 29; mitigation/preparedness procedures, implementation 32; prevention system, water cistern conversion 80–82; preventive system 74; primary hazard 31; primary hazard, mitigation measures *30*; propagation (limitation), vegetation management (usage) 32; response/mitigation strategies, inadequacy 30–31; secondary hazard 31–32
Fire and Rescue Department of Malaysia: GTWHI, collaboration 70; permission, mandatoriness 72
firefighting: infrastructure, insufficiency 75–76; proposals (Gjirokastra, Albania) 80
fire risk: increase 31; mitigation method, application 33; mitigation strategies, implementation 24; monuments exposure, marking 34; Punakha Dzong 110; source 30
Fire Risk Assessment (FRA) 26; methodology, proposal 26, *27*
Fire Risk Mitigation (FRM) 26; measures, proposal 31
'Fire Safety Guidelines for Heritage Buildings' 68
fit-for-purpose index-based methodology 39–40
flood risk (Punakha Dzong) 109
flood risk, identification 88
foundations, involvement 49
Frascaro community: CHIEF ETS, interaction 132–133; Cultural Property Protection Company of the Italian Army, operations control 132; devotion, increase 133
freeze-thaw events, change 46
'From Recovery to Risk Reduction for Sustainability of Historic Areas' 4
future risk, definition 18

Geographical Information system (GIS)-based platform, usage 38, 43, 138
Geographical Information system (GIS) database, establishment 52
Geographical Information system (GIS), spatial analysis 48
George Town UNESCO World Heritage Site, hazards 67
George Town World Heritage City (Malaysia): action plans, implementation 68–72; community-based capacity building 69; community-based Disaster Risk Management 72–73; community-based fire preparedness/response strategy 71–72; cultural heritage, Disaster Risk Management 66, 67–68; cultural heritage, first aid 72; cultural heritage inventory list, backup copies 69; damage assessment, first aid 72; Disaster Risk Management, cultural heritage elements (enhancement) 68; Disaster Risk Reduction, geographic information system 69–70; Emergency Response Team, setup 70–71; monitoring, technology (usage) 70; public awareness campaigns 70
George Town World Heritage Incorporated (GTWHI) 66–68, 137; Emergency Response Team setup 70–71; heritage properties blueprint backup copies, needs/opportunities 69; success 136
Giuliani, Francesca 37
Gjirokastra (Albania) 74–75; cisterns, manual system (usage) 80–81; fire brigade team, discussions 82; firefighting proposals 80; fire prevention system, water

cistern conversion 80–82; high-risk seismic zone 75; historical cisterns, importance 76; hydrants *79*; monument, historical value/importance 77; project, selection criteria 77; surrounding monuments, fire extinguisher system coverage 77; tangible/intangible heritage, preservation 76–77; threats/vulnerabilities, analysis 75–76; water cisterns, final test *81*; water cisterns, map *78*; water cisterns, pilot solution study 76–80; water levels, monitoring 76; water supply, hydrants (relationship) 77
Glacial Lake Outburst Floods (GLOFs), outburst 109
'Good Practices for Disaster Risk Management of Cultural Heritage' 5–6, 74
Grândola Municipality Civil Protection services, project engagement 46
Great East Japan Earthquake and Tsunami (2011), challenges 4
Great Hanshin Awaji (Kobe) Earthquake, long-term recovery 4
Grimaldi, Dulce María 59

Haret al-Sayad, fire 29
Haret el-Madrasa, fire 29
Haret Hosh Kadam, fire 29
'Harmonizing Coordination to Implement Disaster Risk Reduction Strategy' 67–68
hazard: assessment 48; conditions 41; occurrence 64; summaries, usage 18
hazard assessment (MuNDA) **120**
hazard-related factors, implementation 48
hazards/consequences, likelihood (change) 52
heat waves, change 46
Heritage Asset Management (HAM) 37–38, 43
heritage assets value 46
heritage buildings: importance (Punakha Dzong) 107–108; retention, struggle *12*
heritage elements, values (characterisation) 49
heritage facilities, vacant land (usage) 32
heritage models 21
Heritage New Zealand Pouhere Taonga Act 2014 (HNZPT) 12–13; Act National Historic Landmarks/Ngā Manawhenua o Aotearoa me ōna Kōrero Tūturu programme 16
heritage places/ownership/management models, range 16
heritage places, risk identification/management 16
heritage sites: hazardous events, short-term/long-term effects 39–40; providing, manual suppression equipment (usage) 32
heritage values, impact 18
high-rise buildings, construction 25
high-rise towers, walls connection 38
historical buildings: new building construction *26*; stoves, removal 32
Historical Centre (Rethymno, Greece) 45
'Historical Centre of Rethymno' pilot site, STORM platform *53*
historical structures, conservation 88
Historic Cairo: DRMP need 24; monuments/buildings, fire risk (impact) 26; monuments/historical buildings, fire (destructive effect) 29; outstanding universal values (OUV) 25; sites, suffering 24; zone, selection 25
'Historic Cities of the Straits of Malacca, Melaka and George' 66
historic water cisterns, fire preventive system 74
'How to Undertake Integrated Post-Event Damage and Risk Assessment' (ICCROM) 72
Humberstone and Santa Laura Saltpetre Works (Chile) *96*; critical threats 98–99; data collection/analysis 98; Disaster Risk Management (DRM) awareness, creation 101; Disaster Risk Management Plan (DRMP), development initiative 97; Disaster Risk Management Plan (DRMP), implementation 104; Disaster Risk Reduction, activities proposal *102*; Disaster Risk Reduction, stakeholder

articulation proposal *103*; earthquakes/humidity, impact 100–101; inter-institutional working group, establishment/ training 101; risk assessment 99–104; risk management plan 97–98; risk management plan, integration attempts 103; Saltpetre Museum Corporate (CMS), site management 97; site components, risk assessment *100*; vulnerability assessment 103; vulnerability factors 98–99; World Heritage Site, Disaster Risk Management Plan (DRMP) 95
Hurtado, Marcela 95
hydrostatic pressure, increase 41
Hyogo Framework 66

ignition source 29–31; elimination 32; reduction 32
immovable heritage assets, values 46
indicator-based method, application 49
indicator-based vulnerability assessment method 48
informal waste collection places, vacant lands (identification) 33
insignificant, matrix term (usage) 20
Institute of Disaster Mitigation for Urban Cultural Heritage 69
Institute of Disaster Mitigation for Urban Cultural Heritage at Ritsumeikan University (R-DMUCH) 1
institutional coordination 136–137
intangible cultural heritage (Dominican Convent of Santo Domingo Tehuantepec) 88
Intangible Cultural Heritage bill, draft 110–111
intangible heritage routes, protection 88
'Integrated Approaches for Disaster Risk Mitigation of Historic Cities' 4
'Integrated Approach for Movable and Immovable Heritage for Disaster Risk Management of Heritage Sites as well as Museums' 4–5
intense rainfall, defining 48
International Centre for the Study of the Preservation and Restoration of Cultural Property (ICCROM) 2, 72, 74
International Council on Monuments and Sites (ICOMOS) 103
International Disaster Reduction Conference (IDRC), Declaration (adoption) 1, 137
International Training Course (ITC) 1; cultural heritage, DRM theme 2–3; Disaster Risk Management Plan (DRMP), ideation 119; objectives 2; previous courses, theme 3–5; resources 121; risk mapping methodology 61–64
International Training Course (ITC) on Disaster Risk Management of Cultural Heritage 13

Jaho-Babaramo house: cistern, maximum water capacity 81; pilot project selection 80
Je Khenpo 108
Jigyasu, Rohit 1, 135
João Revez, Maria 45
joints, involvement 49

kaitiaki 17
Kim, Dowon 1, 135
Kokalari house, burning 75
Kota Lama, pilot site 66–67

Levuka Historical Port Town (Fiji), pilot site 67
load-bearing walls, involvement 49
local cultural heritage, disaster risk management plan 84

Magliano medieval town walls, collapse 40
major, matrix term (usage) 20
Mamani, Elena 74
Māori, British Crown (treaty) 11
masonry: automatic annotation *42*; wall 41
material susceptibility, reduction 46
medieval town wall (Magliano), collapse 40–41
medieval town wall (Pistoia), collapse 40
medieval walled systems, composition 38
Mellor Heritage Project (UK) 45
Mexican National Agency for Cultural Heritage in Mexico (INAH), conservation project 59–60

Mexico: archaeological decorative elements project 59; Tajín archeological site, collapsed shelter *60*
Ministry of Cultural Heritage and Tourism (MiBACT) 128, 129
minor, matrix term (usage) 20
mitigation measures, proposal 120–121
moderate, matrix term (usage) 20
Monitoraggio delle Mura Urbane (M.O.M.U.) 37
monks, engagement (Punakha Dzong) 112–113
Monte Albán (World Heritage site) 60
monuments, fire risk exposure (marking) 34
Movable Cultural Property Act of Bhutan (2005) 110–111
movable heritage assets, values 46
Mukai, Junko 107
multi-hazard approaches 39
multi-hazard risk: analysis/monitoring/governance 37; approaches, conceptual/methodological framework 39–40; hazards/threats 40–41
multi-leaf masonry 42
Museo Nazionale D'Abruzzo (MuNDA) 119; bureaucracy, role (problems) 126; bureaucratic contingencies, provision 125–126; consultation, recipient feedback (importance) 125; Disaster Risk Management Plan (DRMP), implementation 122; Disaster Risk Management Plan (DRMP), proposal 119–120; Disaster Risk Management (DRM) project, implementation problems 123, 125; Disaster Risk Reduction for Cultural Heritage (DRR4CH) *123*; Disaster Risk Reduction (DRR) knowledge, absence 125; Disaster Risk Reduction (DRR) measures 121; discussion 125–126; earthquake scenario 120; hazard/vulnerability assessment **120**; institution, interaction 121–122; learning points 125–126; low-budget activities, prioritisation 126; mitigation/preparedness measures, proposal 121; MuNDA DRR4CH catalyst 122; pilot project 121; Pilot Project MuNDA DRR4CH catalyst, scope (broadening) 123; preparedness/internal capacity building (enhancement), staff training (prioritisation) 122; presentation/review 122–125; project implementation 121–125; reviewed project, stakeholders (outline) *124*; revised proposal, approval 123, 125; suggestions/requests/customisation 122–123; temporary location, risk assessment (performing) 122–123
Museum of XXI Century Arts (MAXXI) 123

National Agency for Cultural Heritage in Mexico (INAH) 62
National Centre for Disaster Prevention Risk Atlas (CENAPRED): quantitative data 85; satellite images/GIS files, retrieval 86
National Centre of World Heritages Sites (CNSPM) 95; designation 103; process, coordination/conducting 97–98
National Disaster Management Agency, cultural heritage elements 68
National Economic and Development Plan, INAH development 59
National Emergency Management Agency (New Zealand) 16, 137
National Heritage Landmarks legislative requirements 15
National Historic Landmarks/Ngā Manawhenua o Aotearoa me ōna Kōrero Tūturu 13; programme 16
National Institute of Anthropology and History (INAH) 85–86, 137
National School of Conservation, Restoration and Museography (ENCRyM) 87
National Security Council, Directive No. 20 (cultural heritage elements) 68

National Seismic Service earthquake database (Mexico) 84
New Zealand AS/NZS ISO 31000:2009 14–15; risk management process basis *15*
New Zealand Civil Defence and Emergency Management (CDEM) 14
New Zealand heritage places, risk management plans preparation: communication/engagement 17; guideline development 11; guideline framework 16; guideline preparation 15–16; implementation/monitoring/ review 20–21; legislative context 12–13; place/physical context, understanding 18; risk criteria, establishment 17–18; risk evaluation matrix *19*; risk evaluation matrix, terms (usage) 20; risk management policy, development 17–18; risk treatment/management 20; risk, understanding 18; Te Pitowhenua/Waitangi Treaty Grounds, national historic landmark (case study) 13–15; working group, establishment 13
New Zealand risk management framework (alignment), risk management planning (4 Rs methodology) 16
New Zealand Standard, outlines 14
Norcia, Neolithic human settlements (traces) 132

organisational culture, promotion 16
orthoimage, usage 43
outstanding universal values (OUV) 25

paint film, consolidation (San Salvatore a Campi) *132*
pampinas/pampinos 95; continuation 103–104; participation 98
participatory Disaster Risk Management Plan (participatory DRMP), design 85
'Participatory Mapping of Risks for Cultural Heritage Towards the Construction of Resilience and Coresponsibility' 86–87
participatory risk mapping workshop (Dominican Convent of Santo Domingo Tehuantepec) 86–87
Pieve di Santa Maria 130
Pisa, town walls 41–43; analysis, preliminary results *42*
Pistoia, medieval town wall (collapse) 40
planned preventive conservation, implementation 37–38
population activities, monitoring 34
Portuguese General Directorate of Cultural Heritage (DGPC) 46
post-disaster phases 52
primary information collection, training 61
Prince Claus Fund 74, 80
professionals, capacity building 7
project sites, risk map *47*
'Protecting Cultural Heritage From Risks of Natural Disasters Including Those Induced by Climate Change' 4
'Protecting Living Cultural Heritage From Disaster Risks Due to Fire' 4
'Protection of cultural Heritage from Earthquakes and Floods, and Other Associated Hazards' 4
Proyecto de Conservación de Bienes Culturales Muebles Asociados a Inmuebles Arqueológicos en la Región Centro-Sur-Golfo de México (PCSG) 59–60
Punakha District Disaster Management and Contingency Plan (DDMCP), impact 111
Punakha Dzong (Bhutan) *108*; buildings, retrofitting 109; Central Monk Body 108, 113–114; disaster risk 108–110; Disaster Risk Management Plan (DRMP) 107; Disaster Risk Management Plan (DRMP) measures, practice (combination) 113–114; Disaster Risk Management Plan (DRMP), outline 111; Disaster Risk Management Plan (DRMP), preparation 110–111; Division for Conservation of Heritage Sites (DCHS), setup 110; earthquake, risk 109; fire, risk/vulnerability 110, 112; floods, risk 109;

Glacial Lake Outburst Floods (GLOFs), outburst 109; heritage building, importance 107–108; monastic function, sustaining 112; monks, engagement 112–113; potential capacity (enhancement), stakeholders (interactive participation) 112; previous disaster, learning lessons 114–115; project implementation, Disaster Risk Management Plan (DRMP) usage 111–115; risks, understanding 112; Special Commission for Cultural Affairs, establishment 110; Wangduephodrang Dzong fire *114*
Punakha-Wangduephodrang valley, early warning system installation 109

Qasaba street (al-Mu'izz street), conservation/protection 25
qualitative index-based methodologies 39

Ravankhaha, Mohammad 45
real-time shared digital map, generation 64
recipient feedback, consultation importance 125
recovery plan 52
'Reducing Disaster Risks to Historic Urban Areas and Their Territorial Settings Through Mitigation' 4
Regional Directorate of National Monuments, project proposal presentation 77
registered monuments, security (absence) 30
residential places, fire risk (increase) 31
resiliency, capacity building (importance) 72
response procedures, planning 2–3
risk: analysis/evaluation 51–52; avoidance, inclusion 52; criteria, establishment 17–18; evaluation matrix *19*; index, concept *47*; mapping, methodology 61–64; multi-hazard risk analysis/monitoring/governance 37; path, blocking 33; preparedness 52; prevention 52; sources, removal 52; understanding 16, 18, 20
risk assessment 28–29, 45, *101*, 135–136; performing 122–123; process, outline 18; tool 53
Risk Assessment and Management (RA&M) Tool, implementation 53
risk components analysis 51–52
risk management: framework development 16; planning, 4 Rs methodology 16; planning, understanding 16; policy, creation/development 17–18
risk management plan: Humberstone and Santa Laura Saltpetre Works 97–98; preparation guideline, development 11
risk maps *89*; generation 61
risk mitigation 52; strategies, development 45
risk reduction: measures, prioritisation 39; strategies 52
risk treatment 45; guidelines 46
Ritsumeikan University 74; cultural heritage, Disaster Risk Management 5–6; ITC risk mapping methodology 61; UNESCO Chair Programme on Cultural Heritage and Risk Management 13
Roman Ruins (Troía, Portugal) 45
roofs, involvement 49
rubble masonry, inner core 42

Safeguarding Cultural Heritage through Technical and Organisational Resources Management (STORM) *see* STORM
Salah Al-Sharief, Abdelhamid 24
Saltpetre Museum Corporate (CMS), site management 97
San Salvatore (Campi, Norcia), CHIEF ETS intervention 130–132; paint film, consolidation *132*; volunteers, action *131*
Sarris, Apostolos 45
Seh Tek Tong Cheah Kongsi, rescue techniques 71
self-combustibility 31
semi-quantitative ranking, application 48
Sendai Framework on Disaster Risk Reduction 1
Shakya, Lata 1, 135
site layout, benefits 33

site-specific measures, proposal 46
situational awareness Web-GIS service *53*
slow-onset hazards 41
SMS, usage 70
Special Commission for Cultural Affairs, establishment 110
Squad *Pantas,* alert 70
staff training, prioritisation 122
stakeholders: approach 64; articulation proposal *103*; decision-making 7; Dominican Convent of Santo Domingo Tehuantepec analysis 85–86; engagement 136; engagement, promotion 16; interactive participation 112; involvement 32; outline *124*
Standard Operation Plan, usage 69
'Stories of Stones' (RAI Award for best documentary) 133
STORM: CEA methodology 52; context, establishment 45; exposure assessment 49; hazard assessment 48; platform, situational awareness Web-GIS service *53*; project perspective 45; risk analysis/evaluation 51–52; risk assessment/management tool 53; risk assessment procedure 46–52; risk index 46, 48; risk, reduction strategies 52; vulnerability assessment 49, 51
'STORM Classification of Hazards and Climate Change-related Events' 48
stoves, removal 32
streets, fixed/temporary obstacles (determination) 33
Strengthening Disaster Risk Management 95
structural/non-structural vulnerabilities, integration (importance) 136
structural vulnerabilities 88
sudden-onset events, impact 39–40
sudden-onset hazards 46, 52
suppression systems, usage 32
Swedish International Development Agency (SIDA) 74, 80

Tajín (World Heritage site) archeological site, collapsed shelter *60*
Tanner, Vanessa 11
target monuments, identification/map marking 33
Technical University Federico Santa María (USM) 95, 136
technology: innovation 137–138; usage 70
Teotihuacan (World Heritage site) 60
Te Pitowhenua/Waitangi Treaty Grounds, national historic landmark (case study) 13–15
Te Tiriti o Waitangi (Treaty of Waitangi, The) 13–14
threats, analysis 75–76
Torres Castro, David A. 84
town walls: failures 40, *41*; partition, slow-onset hazards (association) *42*
Tróia: earthquake risk map *51*; structural measures 52
Tuscany: ancient town walls, city preservation *38*; historical town walls, damage 37

UNESCO: Chair Programme on Cultural Heritage and Risk Management 13; sustainable development goals 66
UNESCO World Heritage City 71
United Nations Office for Disaster Risk Reduction 28
United Nations World Conference on Disaster Reduction (UN-WCDR) 1
Urban Heritage Site, fire risk mitigation strategies (implementation) 24
urbanism 31

vacant lands, waste collection: places, identification 33; usage 31, 32
Vargas, Mónica 59
vegetation: management, impact 32; presence 40
vertical passing-through cracks, closure 40–41
vulnerability 61, 64; analysis 75–76; assessment 49; factors (Humberstone and Santa Laura Saltpetre Works) 98–99; mapping 88; maps *89*; site information 62
vulnerability assessment (MuNDA) **120**

Waitangi National Trust, COVID-19 awareness 15
Wangduephodrang Dzong fire 110, 113–115, *114*
waste collection sites, identification 34
water cisterns: map *78*; pilot solution study 76–80
water supply, hydrants (relationship) 77
Web-GIS services, spatial analysis 53
wet/dry periods, prolongation 46
wooden/composite cultural heritage, long-term recovery 4
World Heritage Centre 103
World Heritage Sites 136; Disaster Risk Management (DRM) awareness 101; Disaster Risk Management Plan (Humberstone and Santa Laura Saltpetre Works) 95; Manager Office, site management agency 68; project 67; social network 68; workshops 4

Xochicalco (World Heritage site) 6

Zhabdrung Ngawang Namgyal 108
Žuvela-Aloise, Maja 45